REMOTE SENSING OF SHELF SEA HYDRODYNAMICS

Elsevier Oceanography Series, 38

REMOTE SENSING OF SHELF SEA HYDRODYNAMICS

PROCEEDINGS OF THE 15th INTERNATIONAL LIEGE COLLOQUIUM
ON OCEAN HYDRODYNAMICS

Edited by

JACQUES C.J. NIHOUL

Professor of Ocean Hydrodynamics,
University of Liège
Liège, Belgium

ELSEVIER
Amsterdam — Oxford — New York — Tokyo 1984

ELSEVIER SCIENCE PUBLISHERS B.V.
1, Molenwerf,
P.O. Box 211, 1000 AE Amsterdam, The Netherlands

Distribution for the United States and Canada:

ELSEVIER SCIENCE PUBLISHING COMPANY INC.
52, Vanderbilt Avenue
New York, N.Y. 10017, U.S.A.

Library of Congress Cataloging in Publication Data

International Liège Colloquium on Ocean Hydrodynamics
    (15th : 1983)
    Remote sensing of shelf sea hydrodynamics.

    (Elsevier oceanography series ; 38)
    Bibliography: p.
    1. Ocean circulation--Remote sensing--Congresses.
2. Ocean currents--Remote sensing--Congresses.
3. Continental shelf--Remote sensing--Congresses.
I. Nihoul, Jacques C. J. II. Title. III. Series.
GC228.5.I56 1983   551.47   84-1672
ISBN 0-444-42314-1 (U.S.)

ISBN-0-444-42314-1 (Vol. 38)
ISBN 0-444-41623-4 (Series)

Printed in The Netherlands

FOREWORD

    The International Liège Colloquia on Ocean Hydrodynamics
are organized annually. Their topics differ from one year to
another and try to address, as much as possible, recent problems
and incentive new subjects in physical oceanography.

    Assembling a group of active and eminent scientists from
different countries and often different disciplines, they provide
a forum for discussion and foster a mutually beneficial exchange
of information opening on to a survey of major recent discoveries,
essential mechanisms, impelling question-marks and valuable
recommendations for future research.

    The Scientific Organizing Committee and all the participants
wish to express their gratitude to the Belgian Minister of
Education, the National Science Foundation of Belgium, the
University of Liège, the Intergovernmental Oceanographic Commission
and the Division of Marine Sciences (UNESCO) and the Office of
Naval Research for their most valuable support.

    The editor is indebted to Dr. Jamart for his help in editing
the proceedings.

Jacques C.J. NIHOUL

LIST OF PARTICIPANTS

BALLESTER, A., Prof. Dr., Instituto Investigaciones Pesqueras,
    Barcelona, Spain
BÖHM, E., Dr., Dipartamento di Fisica, Universita Roma, Italy.
BOUKARY, S., Mr., University of Niamey, Niger.
CARSTENS, T., Prof. Dr., Norwegian Hydrodynamic Laboratories,
    River and Harbour Laboratory, Trondheim, Norway.
CHABERT D'HIERES, G., Eng., Université Scientifique et Médicale
    de Grenoble, Institut de Mécanique, Grenoble, France.
CLEMENT, F., Mr., Mécanique des Fluides Géophysiques, Université
    de Liège, Belgium.
CREPON, M., Dr., Laboratoire d'Océanographie Physique, Museum
    d'Histoire Naturelle, Paris, France.
DANIELS, J.W., Mr., Department of Oceanography, University of
    Southampton, U.K.
DISTECHE, A., Prof. Dr., Laboratoire d'Océanologie, Université de
    Liège, Belgium.
DJENIDI, S., Eng., Mécanique des Fluides Géophysiques, Université
    de Liège, Belgium.
DUPOUY, C., Miss, Laboratoire d'Optique Atmosphérique, Université
    des Sciences et Techniques de Lille, France.
DYKE, P.P.G., Department of Mathematics and Computer Studies,
    Sunderland Polytechnic, U.K.
GASPAR, Ph., Mr., Institut d'Astronomie et de Géophysique,
    Université Catholique de Louvain, Belgium.
GIDHAGEN, L., Mr., Swedish Meteorological and Hydrological
    Institute, Norrköping, Sweden.
GILLOT, R.H., Dr., Joint Research Centre, Commission of the
    European Communities, Ispra, Italy.
GOFFART, A., Miss, Laboratoire de Biologie Marine, Université de
    Liège, Belgium.
GORDON, C.M., Mr., Naval Research Laboratory, Washington, U.S.A.
GOWER, J.F.R., Dr., Institute of Ocean Sciences, Sidney, Canada.
GRILLI, S., Hydraulique Générale et Mécanique des Fluides,
    Université de Liège, Belgium.
GROSJEAN, P., Mr., Mécanique des Fluides Géophysiques, Université
    de Liège, Belgium.
HECQ, J.H., Dr., Laboratoire de Biologie Marine, Université de
    Liège, Belgium.

JACOBS, W., Mr., Institut für Geophysik und Meteorologie der
    Universität Köln, Germany.

JAMART, B., Dr., Unité de Gestion, Modèle Mathématique Mer du
    Nord et Estuaire de l'Escaut, Cellule de Liège, Belgium.

LEBON, G., Prof., Dr., Thermodynamique des Phénomènes Irréversibles
    Université de Liège, Belgium.

LE CANN, B., Mr., Laboratoire d'Océanographie Physique, Université
    de Bretagne Occidentale, Brest, France.

LIN, S., Mr., The Second Institute of Oceanography, Hangchow,
    Zhejiang, People's Republic of China.

LOFFET, A., Eng., Belfotop Eurosense, Wemmel, Belgium.

LYGRE, A., Mr., Continental Shelf Institute, Trondheim, Norway.

MARUYASU, T., Prof., Dr., The Science University of Tokyo, Noda,
    Chiba, Japan.

MASSIN, J.M., Dr., Ministère de l'Environnement, Direction de la
    Prévention des Pollutions, Neuilly, France.

MONREAL, M.A., Mrs., Consejo Nacional de Ciencia y Tecnologia
    (Conacyt), Mexico.

MORCOS, S., Dr., Division des Sciences de la Mer, UNESCO, Paris,
    France.

MURALIKRISHNA, I.V., Dr., National Remote Sensing Agency,
    Balanagar, India.

NEVES, R., Mr., Instituto Superior Tecnico, Lisboa, Portugal.

NIHOUL, J.C.J., Prof., Dr., Mécanique des Fluides Géophysiques,
    Université de Liège, Belgium.

NISHIMURA, T., Dr., The Science University of Tokyo, Noda, Chiba,
    Japan.

ONISHI, S., Prof., Dr., The Science University of Tokyo, Noda,
    Chiba, Japan.

PIAU, P., Eng. Institut Français du Pétrole, Rueil-Malmaison,
    France

PINGREE, R.D., Dr., Marine Biological Association, Plymouth, U.K.

POULAIN, P.M., Mr., Mécanique des Fluides Géophysiques, Université
    de Liège, Belgium.

RONDAY, F.C., Dr., Mécanique des Fluides Géophysiques, Université
    de Liège, Belgium.

SALAS DE LEON, D., Mr., Consejo Nacional de Ciencia y Tecnologia
    (Conacyt), Mexico.

SALUSTI, S.E., Dr., Istituto di Fisica, Universita Roma, Italy.

SENCER, Y., Eng., Mithatpasa cad, Ankara, Turkey.

SMITZ, J., Eng., Mécanique des Fluides Géophysiques, Université
de Liège, Belgium.

TANAKA, S., Dr., Remote Sensing Technology Center, Tokyo, Japan.

VENN, J.F., Mr., Mathematics Department, City of London
Polytechnic, U.K.

VAN DER RIJST, H., Dr., Elsevier Publ. Company, Amsterdam,
Holland.

WITTING, J.M., Dr., Naval Research Laboratory, Computational
Physics, Washington, U.S.A.

YENTSCH, C.S., Prof., Dr., Bigelow Laboratory for Ocean Sciences,
Maine, U.S.A.

CONTENTS

WATER COLOUR IMAGING FROM SPACE

J.F.R. GOWER

Institute of Ocean Sciences, P.O. Box 6000, Sidney, B.C., Canada
V8L 4B2

ABSTRACT

Water colour images from the Coastal Zone Color Scanner on the
NIMBUS 7 satellite can now show physical and biological processes
in the ocean with greater clarity than has ever been possible
before. Examples are presented here of turbulent flow patterns
in the Gulf Stream affected by the New England seamounts, coastal
upwelling off South Africa, the surface pattern formed by the
Alaskan Stream, and regions of high phytoplankton concentration
on the continental shelf of Argentina. The processing steps now
being used to obtain these results are described, with references
to more detailed treatments. Possibilities for future
improvements in this type of remote sensing measurement are
discussed, with particular reference to the possibility of
mapping naturally stimulated phytoplankton pigment fluorescence
from space.

INTRODUCTION

The colour of the sea has been used by sailors for centuries
as a check on navigation and as an aid to locating productive
waters for fishing. Several seas round the world are named after
their colours, the most commonly cited example being the Red Sea,
named after its sporadic blooms of the phytoplankton
Trichodesmium (= Oscillatoria). Currents bring together water
masses with more subtle colour differences. The Kuroshio ("dark
water") is named for this difference, and the colour change at
the edge of the Gulf Stream can also be distinguished by eye from
a ship.

Near the coasts the colour changes can be due to resuspension
of bottom sediments in shallow water or to river discharge of
silt-laden water. Water from the Yangtse Kiang river in China,
for example, gives the Yellow Sea its name. Colour changes
further from shore must be due to the growth of phytoplankton
where conditions of nutrients and sunlight are favorable. Such

growth can cause colour changes from blue through blue-green to green and in extreme cases to yellow, brown or red. Patches and streaks of strongly discoloured water were widely reported by early travelers. Darwin (1845) for example, cites references and describes passing through large areas off South America where "the colour of the water as seen at some distance was like that of a river which has flowed through a red clay district; but under the shade of the vessel's side it was quite as dark as chocolate. The line where the red and blue water joined was distinctly defined. The weather for some days previously had been calm and the ocean abounded to an unusual degree with living creatures." Darwin here points out two important connections--with calm weather, allowing near surface stratification, stability and high growth rates, and with the other "living creatures" in the ocean who depend on phytoplankton as the first link in their food chain.

Early observers were greatly intrigued by the fronts and narrow bands exhibited by the patches of visibly discoloured water. In fact only minute elements of the full patterns can be seen from a ship. The satellite images presented below show more of the full complexity of structure due to current streams, and mesoscale eddy fields influenced by larger scale water movements.

The water colour images are from the Coastal Zone Color Scanner on NASA's NIMBUS 7 satellite. Processing and correction techniques used in deriving these images are discussed. Opportunities for future developments are suggested, including the possibility of mapping naturally stimulated phytoplankton pigment fluorescence from satellites.

SATELLITE WATER COLOUR IMAGERY

Early weather satellites provided visible and thermal images of clouds, to show the locations of weather systems by day or night. The thermal imagery could also faintly distinguish the sea surface temperature structure associated with major boundary currents such as the Gulf Stream. Improvement of these sensors, through the SR (Scanning Radiometer) to the VHRR (Very High Resolution Radiometer) to the present AVHRR (Advanced Very High Resolution Radiometer) now gives clear and sharp images, such as Fig. 1, which can resolve temperature and spatial differences as small as 0.2°C and 1 km respectively. This image shows the thermal patterns of the surface skin of the ocean, associated with the northeastward flow of the Gulf Stream off the coast of North

Fig. 1. Thermal infrared image (TIROS N AVHRR) showing the Gulf Stream at 19.22 GMT on May 7, 1979.

America on May 7, 1979. Warmer water is dark, and cold, high clouds appear white. Eddies shed by the stream can be seen to the north and south of the main current.

The degree of detail and the geometrical fidelity of these images have made them a major tool of physical oceanography. By contrast, the associated visible images (Fig. 2) have been useful only indirectly, providing additional information on the presence of low cloud in daytime passes, though in some cases ocean information can also be deduced from sunglint patterns (La Violette et al, 1980). In Fig. 2 the only contrast visible over the water are the white patches due to cloud, with a faint brightening at the lower left due to sunglint.

The amount of light upwelling from beneath the sea surface gives only about 1% of the signal from sunlit clouds, so that variations in this quantity should indeed be hard to detect on an image designed for cloud mapping. Thermal contrast due to the Gulf Stream, on the other hand, can easily amount to 5% of the full scale signal, making this an easier target for satellite remote sensing.

4

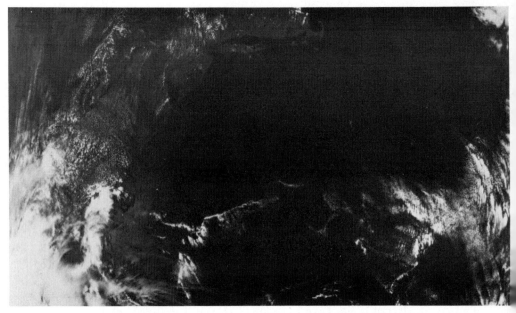

Fig. 2. Visible image recorded at the same time as Fig. 1.

A specialized ocean colour sensor can, however, do much better than Figure 2 would suggest. Sensitivity can be increased and the signal allowed to saturate over cloud. A mirror can be used to tilt the field of views away from areas where sunglint is expected, and narrow, optimally placed spectral bands can be selected. The result of this improvement is dramatically illustrated in Fig. 3 which shows processed data from the Coastal Zone Color Scanner on NIMBUS 7, for the same area of ocean at nearly the same time (3.5 hours earlier).

Shades of grey in the image represent phytoplankton chlorophyll a and phaeophytin pigment concentrations (a standard measure of phytoplankton concentration) with the darkest shades corresponding to .05 mg.m^{-3} and the lightest to over 10 mg.m^{-3}. Comparing this with Fig. 1, the water colour image is able to show more structure in the water, though Fig. 1 could possibly be further enhanced to bring out more structure in the colder water. The features that appear on the two figures are very similar, illustrating the commonly observed high anticorrelation between temperature and phytoplankton concentration. The Gulf Stream is again darker in Fig. 3, but

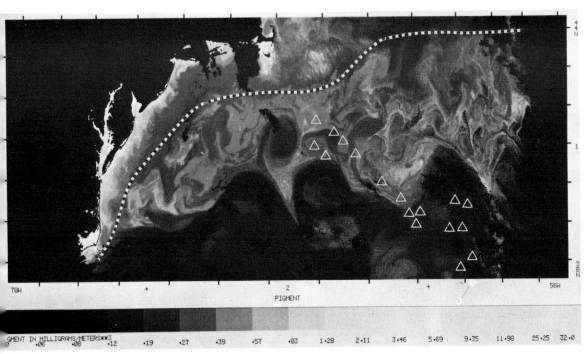

Fig. 3. CZCS processed (level 2) pigment image showing the same area as Fig. 1 at 1554 GMT on May 7, 1979 (3.5 hours earlier) 1000 m depth contour (dotted) and New England seamount chain (triangles) have been superposed. Grey tone step wedge, below gives pigment value.

here because of its low pigment concentration (< 0.1 mg.m^{-3}), compared to more than 0.3 mg.m^{-3} in the more productive waters further north.

The complex mesoscale eddy field in the water north of the Gulf Stream is here well illustrated. Such spatial patterns can be quantized in terms of their two dimensional spatial spectrum (Gower et al. 1981) and variations in this spectrum, such as those suggested by the increase in high frequency structure on the right side of the image, would be expected to correlate with the changing dynamics of different ocean areas. In the present case, the chain of New England seamounts (plotted as triangles) crosses the image where this change in structure is observed. The seamounts extend up from the bottom at 5000 meters to depths of between 1000 and 2000 meters, and therefore intercept the Gulf Stream, which flows in the top 2500 meters of the ocean. Richardson (1981) has reported the effect of these seamounts on the surface flow as traced by buoy tracks. He observed meanders and small (20 km scale) eddies, near and extending eastwards from individual seamounts. Fig. 3 illustrates this effect more fully, with several instances of smaller eddies near the seamounts, and a generally more disordered flow to the east.

The 1000 meter depth contour is superposed on the image to show the position of the edge of the continental shelf. The area of high productivity caused by tidal mixing over Georges Bank can be seen as a lighter toned area east of Cape Cod. Fine structure in this area follows the form of the shallow (about 10 m) shoals on the bank.

Fig. 4 shows an area off the west coast of South Africa. Cape Town is at the lower right corner of the image. The light areas indicate high productivity due to coastal upwelling where pigment concentrations can reach 30 mg.m^{-3}. CZCS images of these

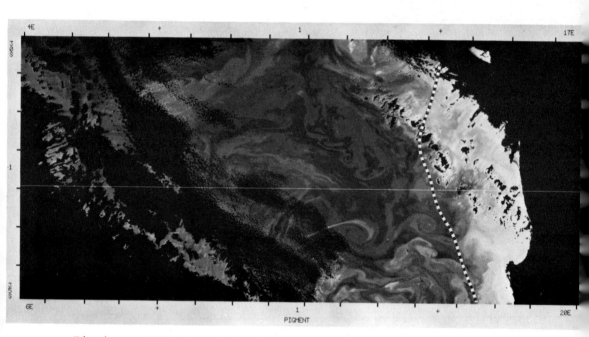

Fig. 4. CZCS pigment image showing effects of upwelling (right) and possibly the Benguela current (bottom right), off South Africa on November 3, 1979. 1000 m depth contour (dotted) superposed.

coastal areas have been discussed by Shannon et al (1983). The change in spatial structure further offshore, at the lower right of the image, again suggests a dynamic input, here from the Benguela current.

Fig. 5 covers a small strip of the north east Pacific Ocean along the southern edge of the Aleutian Island chain, and shows the surface structure associated with the Alaskan Stream on July 10, 1979 in terms of pigment level variations in the range 0.4 to 1 mg.m^{-3}. This stream is the major current by which water leaves the Gulf of Alaska. It flows westward as a narrow jet along the continental slope, whose landward edge is indicated in Fig. 5 by the 1000 m contour (dotted). The structure visible in the image confirms the narrow (60 km) width deduced by Royer (1981) from current meter observations, and shows the start of a recirculating eddy near the bottom centre of the image, south of Dutch Harbor on Unalaska Island. Such recirculation, in which Gulf of Alaska water mixes into the Pacific Ocean, has been observed to occur at a range of longitudes as discussed by Thomson (1972). The position of this eddy is in the start of this range and near the position for the start of recirculation reported by Wright (1981) for March 1980.

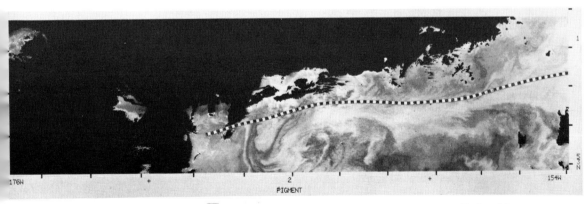

Fig. 5. CZCS pigment image showing the Alaskan stream on July 10, 1979 with the 1000 m depth contour (dotted) superposed.

Fig. 6 shows high concentrations of phytoplankton along the edge of the continental shelf off the Argentina coast, between 40 and 45° South on December 10, 1978. Productivity here is related to mixing by strong tidal currents over the shelf. Areas of high

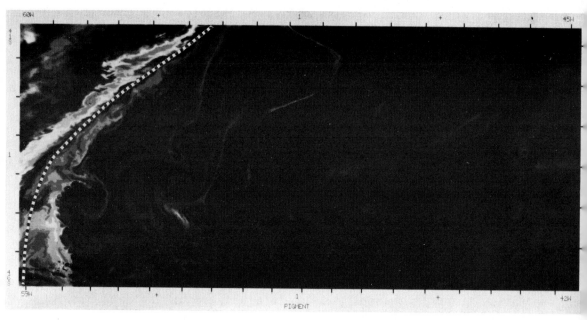

Fig. 6. CZCS pigment image showing areas of high phytoplankton concentration on the edge of the continental shelf off Argentina on December 10, 1978. 1000 m depth contour (dotted) superposed.

phytoplankton concentration have been elongated by current shear parallel to the coast, and further strips of pigmented water are visible further offshore. Similar strips must have led to reports by Darwin and others of "great bands" of discoloured water.

Location of depth contours and seamounts on figures presented above makes use of latitude and longitude marks provided round the edges of processed images. Inaccuracies of about 30 km in positions of these marks were noted in Figs. 3 and 5 and partially corrected by reference to visible coast features. No such correction is possible in Fig. 6 and the depth contour may therefore be mis-located by a similar distance.

In all these figures the data is processed so that grey shades of the image will correspond to definite levels of phytoplankton chlorophyll a and phaeophytin pigment concentration as indicated by the grey wedge shown under Fig. 3. All land and cloud, observed as being above a given near infrared brightness threshold are masked to black so as to suppress grey tones for which this correspondence will certainly not apply.

The form of the colour change being detected in these images is shown in Fig. 7 (NASA, 1982). Low concentrations of pigment will absorb blue light at wavelengths shorter than 500 nm, leading to a change from a blue to a blue/green colour for the water. At higher pigment concentrations backscatter from the associated cellular material in the water increases the radiance observed at longer wavelengths as indicated, leading to a yellow or brown colouration.

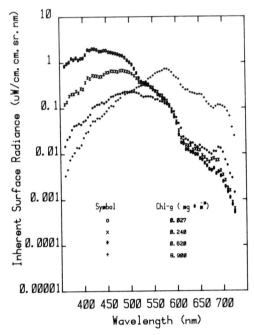

Fig. 7. Sea-water leaving radiance spectra for several chlorophyll a pigment concentrations. (NASA, 1982)

The algorithms used in the processing are accurate only in so called case 1 water where phytoplankton and their covarying detrital material play the dominant role in determining the optical properties (Morel and Gordon, 1980). This is true in open ocean and many coastal areas. In other areas (case 2 water) suspended material from a shallow bottom, or dissolved or suspended material from land will be important. In clear shallow water light reflected from the bottom will also form part of the optical signal.

Fig. 8 shows a part of the Gulf of Mexico and Grand Bahama Bank area where most of the grey shades are due to added radiance reflected from the ocean bottom through the very clear waters where pigment levels are typically near the instrument's detection limit. In most cases this will not be a severe source of error, but it indicates the variety of optical problems encountered by an ocean colour scanner. In this case an algorithm that interpreted observed optical radiances in terms of water depth and bottom reflectance (Lyzenga 1981), might well produce useful results.

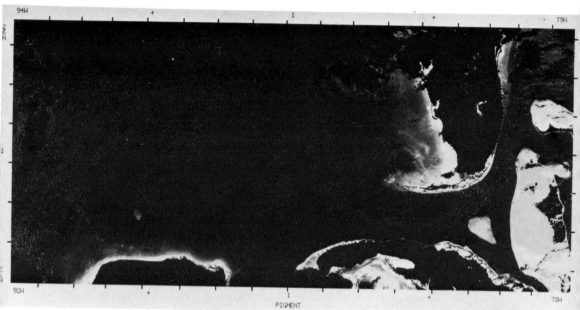

Fig. 8. CZCS pigment image showing shallow water areas in the Gulf of Mexico and on Grand Bahama Bank on December 2, 1978.

PROCESSING OF CZCS WATER COLOUR IMAGES

Water colour data is collected by the CZCS in 4 bands 20 nm wide centred at 443 (blue), 520 (blue/green), 550 (green) and 670 nm (red). A further two bands at 750 nm and 11 μm are used for masking cloud or land and for providing simultaneous thermal images respectively. The thermal band operated only intermittently and ceased working in 1981. Its output was of lower quality than the AVHRR, which can, in principle, provide data with a sufficiently small time difference to make the thermal channel on the CZCS of limited use.

Pigment concentrations and attenuation coefficients are computed using algorithms based on observed correlations of these quantitites with upwelling radiances from the ocean in the blue and green spectral regions (Clark, 1981). To deduce these radiances from CZCS data, the outputs of the first three bands need to be corrected for atmospheric and surface effects. The fourth band at 670 nm is used in making this correction as described below.

The processing of CZCS images as carried out by NASA provides two levels of output (Hovis et al 1980, Hovis 1981). Level 1 gives a set of "quicklook" images of the data in each band recorded by the satellite, and level 2 gives images of: computed sub-surface radiances, corrected for atmospheric and surface effects; the aerosol signal at 670 nm; the phytoplankton pigment concentration; the diffuse attenuation coefficient and the thermal radiance where this band was operating. Grey levels on the level 2 images relate to quantitative values of all these variables. Figs. 3-6 and 8 above are examples of level 2 pigment images.

A number of papers have been published describing improvements that have been made in arriving at the present process, the most recent being Gordon et al (1983). The first step is to convert the measured signals into radiance units. This step has been complicated by a degradation of the reflection of the sensor tilt mirror while in orbit. This is not monitored by the on-board sensor calibrations, but can be accurately followed by its effects on the resulting data, and time dependent calibrations have now been derived. Modifying the calibration for each band in this way will also compensate for errors in the assumed solar spectrum.

The major computation in the processing is to remove the signal due to Rayleigh scattering of sunlight in the atmosphere over the slightly reflecting ocean, with allowance for ozone absorption in the upper atmosphere. Gordon et al (1983) have found that a single scattering approximation works well, but the computation must be carried out for the rather complex geometry of the sensor scan, about an axis tilted to avoid sunglint, over a curved earth. Since the signal depends on the total of gases in the atmosphere, it can be predicted fairly accurately, giving a well defined problem easily handled by computer software. The signal varies smoothly across the scene and can be interpolated after relatively few computations.

The Rayleigh signal and upper atmosphere ozone concentration have a slight seasonal and latitudinal dependence that is allowed for in five possible steps. Variations in atmospheric pressure, and in the surface water reflection with wind and waves, including foam cover, cannot be compensated without more data. Some correction is provided in the next stage of aerosol correction.

Aerosol scattering in the atmosphere adds a signal which is much more variable in intensity, but which has a smooth spectrum which can be reasonably well approximated by a power law. At 670 nm the water radiance becomes very small, and the remaining signal after Rayleigh correction can be used as a measure of the varying aerosol signal in the scene at this wavelength. Extrapolation to the wavelengths of other bands, however, requires a knowledge of the exponent of the power law spectrum. In retrospect, at least two bands, at 670 nm and at a longer wavelength, would have been useful for measuring this exponent at each pixel. The 750 nm band included in the CZCS is of low sensitivity and is not suitable for this purpose. However Gordon (1981) showed that this exponent was often constant over large areas and Gordon and Clark (1981) proposed the currently used method of determining it from one "clear water" point in the scene, and applying the resulting aerosol spectrum to the whole image. The method makes use of the fact that the upwelling radiance from case 1 water containing phytoplankton at a pigment concentration of less than 0.25 mg.m^{-3} and no other significant scattering material, will be close to fixed values at 520 and 550 nm and will be very low at 670 nm (see Fig. 7). The mean aerosol spectrum power law deduced from these three wavelengths can then be extrapolated to 443 nm.

The resulting "aerosol" correction can contain contributions from improperly corrected Rayleigh radiance, surface foam and residual calibration errors, and will tend to reduce these effects where conditions are the same as at the "clear water" point. The correction will be less perfect in other areas of the scene especially if the aerosol properties change. Errors will also exist wherever any suspended material raises the water leaving radiance at 670 nm, since the signal in this band is used to map the varying aerosol contribution whatever its spectrum. An iterative process in which deduced pigment concentrations are used to estimate the 670 nm radiance due to higher concentrations of phytoplankton was proposed by Smith and Wilson (1981), but this has not been implemented in the standard NASA process.

The final stage of the atmospheric correction consists of computing the subsurface radiances that will give observed, corrected satellite radiances. This must allow for the facts that surface refraction reduces the signal from beneath the water by about half, and that Rayleigh scattering and ozone absorption attenuate the signal passing out through the atmosphere.

These subsurface radiances are then used as inputs to the pigment and attenuation coefficient algorithms. Since these algorithms are based on observations in case 1 waters where optical properties are determined by phytoplankton concentration only, the two outputs are in fact highly correlated. The algorithms are in the form of mean power law relations with ratios of subsurface radiances in bands 1 and 3, and 2 and 3 as given by Gordon et al (1983) and SASC (1983). These two documents give details of most of the above processes, with the exception of the time dependent calibration, which is still being refined, and the method used to automatically select clear water areas.

The above processing system seems to work well in that images are produced in which atmospheric aerosol patterns are largely suppressed. A limited evaluation given by Gordon et al (1983) shows that pigment concentration estimates can be accurate to ±30%. However this is for scenes showing large clear areas containing good "clear water" reference areas, and refers only to the pigment concentration range 0 to 1.5 mg.m^{-3}. Shannon et al (1983) studying CZCS images of the relatively cloud free Southern Benguela current region (Fig. 4) find differences between ship and satellite chlorophyll $\underline{a}$ pigment estimates over the range 0.1 to 20 mg.m^{-3}, of about a factor of 2. In many areas the observations will need to be made in smaller clear areas among cloud. Here the existence of good clear water reference areas becomes particularly critical. Gordon et al (1983) show that the effect of only 0.27 mg.m^{-3} of pigment in the "clear water" area can lead to a factor of 2 error in deduced pigment for other areas. A drawback of the present processing system is that the position of the assumed clear water pixel is not recorded on the final data, so that users cannot easily assess possible errors.

It must be emphasized that the present problem in making aerosol corrections is largely due to the present design of the

CZCS. Morel and Gordon (1980) proposed an improved set of spectral bands, since refined in the MAREX report (NASA, 1982), which would greatly reduce this problem.

The examples shown above demonstrate the value of the data. The MAREX report (NASA, 1982) suggests how an improved, follow-on sensor could be used in a large scale program of primary productivity mapping with applications in fisheries, climate studies and physical oceanography.

TECHNICAL IMPROVEMENTS POSSIBLE FOR SATELLITE WATER COLOUR MEASUREMENTS

Improvements which can increase accuracy and coverage of satellite water colour data have been mentioned above and by Morel and Gordon (1980), NASA (1982) and SCOR (1983). Table 1 summarizes these proposals, several of which are being implemented on the next Ocean Color Imager due to be launched by the U.S. by about 1986 on one of the NOAA weather satellites.

A further major problem found with the CZCS was in the complexity of the required data processing, and the resulting long delays before data became available. The problems now seem to have been overcome and the data backlog, in some cases extending back five years, is now being reduced.

Technical developments in the field of integrating optics with solid state electronics have resulted in sensor arrays that can be used for remote sensing, either in a pushbroom mode (where a one-dimensional line of sensors looks at contiguous points along a line of view which is moved at right angles to the line by motion of the satellite) or in an imaging spectrometer mode (where a two-dimensional array of sensors operates as many pushbroom scanners, each at a different wavelength). Such sensor arrays offer high sensitivity and the possibility of observing in more, or more precisely chosen, spectral bands.

A typical sensor array might have 300 by 300 elements, which would allow pushbroom imaging of a 15° field of view with an angular resolution comparable to the CZCS, and a spectral resolution of 1.5 nm in the wavelength range 400 to 850 nm. Several arrays would be required to cover the wider CZCS field of view. If the outputs from all elements were read and digitized at the rate required for satellite imaging (about 10 times per second) then the volume of data would be enormous (about 50 times

TABLE 1

Suggested improvements in satellite ocean colour imagers (OCI)

Improvement	Technical requirement	Current action
Improve aerosol correction	add infrared bands	include in next OCI
Improve pigment characterization	add visible bands	include in next OCI
Map smaller pigment changes	increase sensitivity	include in next OCI
Improve area coverage	add onboard processing to reduce data volume	include in next OCI
	add colour sensors to geosynchronous satellites	proposed but not yet implemented
Distinguish yellow substance from phytoplankton pigments	add band near 400 nm	proposed but not implemented
Map natural fluorescence	increase sensitivity and add special bands	flexible airborne sensor being constructed

that from the present CZCS). The outputs can however be digitally combined into predetermined spectral bands thus reducing the data band width to that required by mechanical scanners, and giving much greater flexibility and precision in selection of the bands. The selection can be changed under software control, allowing a variety of specialized band combinations to be formed for mapping different target signatures. This type of sensor is particularly suitable for attempting the mapping of naturally stimulated phytoplankton pigment fluorescence as discussed in the next section.

MAPPING OF NATURALLY STIMULATED PHYTOPLANKTON, CHLOROPHYLL A FLUORESCENCE IN SEA WATER

The broad band colour changes that are mapped by the CZCS are caused by a combination of absorption and backscattering of incident light by phytoplankton. The resulting colour changes are illustrated in Fig. 7 and are often adequately characterized by green to blue ratios deduced from measurements in CZCS bands.

Another familiar feature of phytoplankton chlorophyll a pigments is their fluorescence, which for the most commonly occurring process leads to emission at 685 nm. A slight increase in the radiance at this wavelength, due to natural stimulation of this fluorescence by sunlight, can be seen for all four spectra plotted in Fig. 7, where the amount of this increase, above a smooth baseline, is roughly proportional to the chlorophyll concentrations listed.

Use of this signal for airborne remote sensing surveys was first suggested by Neville and Gower (1977) and Gower (1980), and for satellite observations by Gower and Borstad (1981). The fluorescence signal has been found to be proportional to the chlorophyll a concentration, though the value of the proportionality constant has been found to vary in the case 2 waters where most tests have been made (Fig. 9). Observations of naturally stimulated fluorescence have been used successfully in airborne surveys of the British Columbia coast (Borstad et al, 1980) and in the eastern Canadian Arctic (Borstad and Gower, 1983). Gower and Lin (1983) report a characteristic vector analysis of reflectance spectra for coastal waters for which fluorescence appears to provide superior estimates of pigment concentrations compared to the estimates derived from green to blue ratios. This analysis has been extended to examine variations in the fluorescence emission for different phytoplankton (Lin, et al, this volume).

An 8 band ocean colour scanner with a band centred at 685 nm, 23 nm wide, was also flown on the Space Shuttle in 1981 (Kim et al, 1982). Other similar bands at 655 and 787 can be used to interpolate a baseline from which the radiance difference at 685 nm may be related to chlorophyll a fluorescence. Unfortunately, apart from the low sensitivity of the sensor and the non-optimal widths and positions of the bands, there were problems with weather and timing of this shuttle flight. The best scene of the limited resulting data set is shown in Fig. 10

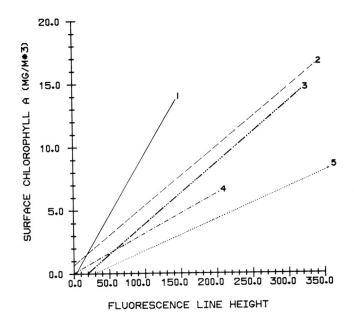

Fig. 9. Relations between naturally stimulated fluorescence (expressed as apparent reflectance increase at 685 nm x 10^5) and phytoplankton pigment concentrations observed in surveys on the British Columbia coast in 1979 (1 and 2), 1981 (3 and 4) and 1976 (5).

with uncorrected radiance at 685 nm (top) and the calculated radiance difference at 685 nm from the linear baseline (bottom). The scene shows parts of the Yellow Sea and the East China Sea off the mouth of the Yangtze River so that most of the water colour changes will be related to suspended sediment. Some of the brightening in the lower image, for example near the coast of Korea (top right), may be due to pigment fluorescence. The sensitivity is such that fluorescence due to a few mg.m^{-3} of pigment should be detectable. The lower scene is much less affected by the aerosol change near Cheja Island (centre) and by the strong limb brightening both visible on the top image. This data has not yet been processed using the techniques described above.

Apart from its use as an estimator of chlorophyll _a_ concentration, the fluorescence signal will provide another tracer of water flow, or mixing patterns. Fig. 11 shows a variation in the observed fluorescence signal between spectra (A

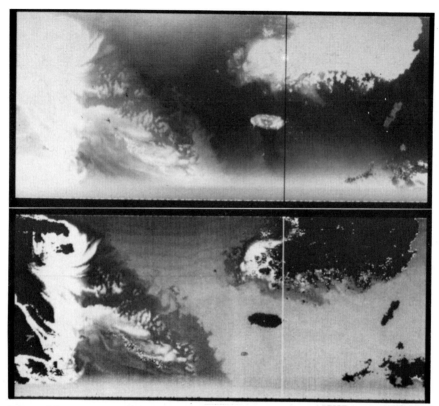

Fig. 10. Images from the OCS experiment on the OSTA-1 Space
Shuttle flight on November 13, 1981, showing the mouth of the
Yangtze River (left) and southern Korea (top right). Uncorrected
685 nm band (top), partly processed fluorescence image (bottom).

and B) taken a few minutes apart in Kiel harbour (Gower,
unpublished). Curve C is the difference plotted with 20 times
more sensitivity. The proportional change at 550 nm is much
smaller than that in the fluorescence signal.

For airborne and satellite remote sensing the fluorescence
signal has the advantages of a narrow band width, which
distinguishes it from the variable, broad band signals due to
aerosols or water surface effects, and a position at the red end
of the optical spectrum where the Rayleigh scattered radiance is
low. Absorption of light by the atmosphere occurs at wavelengths
close to that of the fluorescence signal, particularly on the
longer wavelength side where water vapour absorbs with varying
strength from 690 to 745 nm and oxygen from 687 to 694 and from
760 to 770 (Fig. 12).

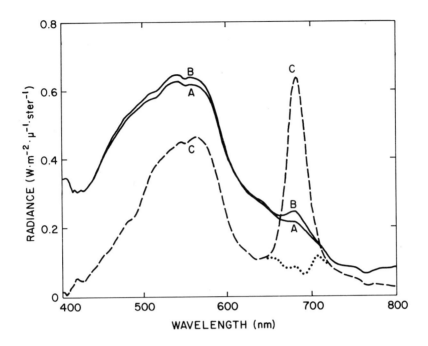

Fig. 11. Water radiance spectra (A and B) observed at two points in Kiel Harbour on April 26, 1982 from the deck of a ship. The difference (C) is plotted at 20 times the vertical scale. The right hand peak in curve C, interpreted as caused by a change in chlorophyll a fluorescence, can be fitted by a Gaussian centred at 682 nm with a half height width of 24 nm (residual shown dotted). Spectrometer resolution is 12 nm.

Observing bands will need to be fitted between these features with the relatively high precision of a few nanometers. Measurement of the fluorescence signal will be by analysing the radiance spectrum shape in the range 660 to 690 nm supplemented by measurements in the window at 745 to 760 nm, or in the almost transparent window at 708 to 714 (Fig. 12) to remove the smoother shape of the background radiance.

Although such observations could be made with a specially configured mechanical scanner, an array sensor such as described above provides greater sensitivity and flexibility. Such a sensor, the Fluorescence Line Imager (FLI), is now being built as part of the remote sensing program of the Canadian Department of Fisheries and Oceans. This is an airborne prototype imaging

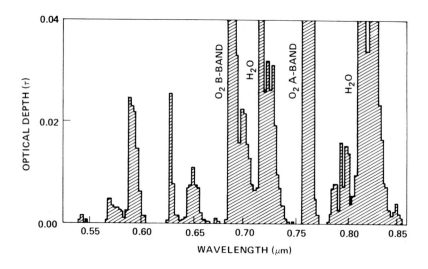

Fig. 12. Atmospheric optical depths between 500 and 850 nm due to absorption by oxygen and water vapour. Note the expanded vertical scale which shows faint features especially at wavelengths shorter than 680 nm.

spectrometer whose properties are listed briefly in Table 2.

Figure 13 shows the sensor head with four of its five cameras, which will together cover a 70° field of view. Fig. 14 shows the layout of one of the cameras in which light is dispersed by a transmission grating and focussed onto the CCD array on the left side. Some of the readout electronics is also visible. Computer control will allow spectral band specification and will perform the processing needed to form these bands by signal summation. A real-time output is available for display of a mathematical combination of different bands.

Flight programs are being planned to test use of this sensor over a variety of targets. Although the instrument was designed specifically for water colour observations, its parameters make it ideal for other remote sensing studies, for example in the fields of agriculture, forestry, geology and atmospheric sciences and for simulating the spectral responses of other optical imagers. Scientists interested in joint observing programs should contact the author.

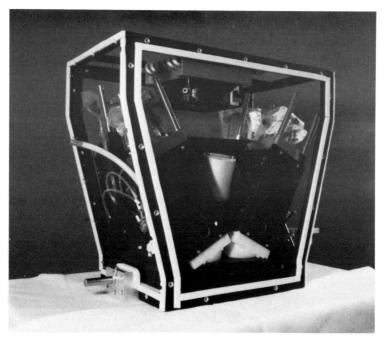

Fig. 13. The sensor head of the Fluorescence Line Imager (FLI), being built for the Canadian Department of Fisheries and Oceans, with four of the five CCD cameras in position.

Fig. 14. One of the FLI cameras with covers removed, showing the layout of the optics and some of the digitizing electronics.

TABLE 2

Properties of Fluorescence Line Imager (FLI)

Size of arrays used	385 x 288
Number of arrays	5
Total field of view	70°
Total number of pixels	1925
Total number of spectral elements	288
Spectral coverage	410 to 850 nm
Spectral resolution	2 nm
Number of bands	8
Location and width of bands	under software control to 1.5 nm
Digitization	12 bits
Signal to noise	2000:1 for a 30 nm band
Scan rate	10 per second

CONCLUSIONS

Processed CZCS imagery demonstrates the potential of ocean colour imaging from space for physical as well as biological oceanography. Improved sensors should lead to more precise results, covering wider areas with greater regularity. Imaging of natural fluorescence also appears possible and should lead to further improvements.

REFERENCES

Borstad, G.A., Brown, R.M., and Gower, J.F.R., 1980. Airborne remote sensing of sea surface chlorophyll and temperature along the outer British Columbia coast. Proceedings 6th Canadian Symposium on Remote Sensing, Halifax, N.S., May, pp. 541-549.

Borstad, G.A. and Gower, J.F.R., 1983. Ship and aircraft measurements of phytoplankton chlorophyll distribution in the eastern Canadian Arctic. Arctic, in press.

Clark, D.K., 1981. Phytoplankton pigment algorithms for the NIMBUS-7 CZCS. In: J.F.R. Gower (Editor), Oceanography from Space. Plenum Press, Marine Science, 13: 227-238. New York.

Darwin, C.R., 1845. The voyage of the Beagle, 2nd Ed., Everyman Library Paperback, Dent, London.

Gordon, H.R., 1981. A preliminary assessment of the NIMBUS-7 CZCS atmospheric correction algorithm in a horizontally inhomogeneous atmosphere. In: J.F.R. Gower (Editor), Oceanography from Space, Marine Science 13: 257-265. Plenum Press, New York.

Gordon, H.R. and Clark, D.K., 1981. Clear water radiances for atmospheric correction of coastal zone color scanner imagery. Applied Optics, 20: 4175-4180.

Gordon, H.R., Clark, D.K., Brown, J.W., Brown, O.B., Evans, R.H. and Broenkow, 1983. Phytoplankton pigment concentrations in the Middle Atlantic Bight: Comparison of ship determinations and CZCS estimates. Applied Optics, 22: 20-37.

Gower, J.F.R., 1980. Observations of in situ fluorescence of chlorophyll a in Saanich Inlet. Boundary Layer Meteorology, 18: 235-245.

Gower, J.F.R., Denman, K.L. and Holyer, R.J., 1980. Phytoplankton patchiness indicates the fluctuation spectrum of mesoscale turbulence. Nature, 288: 157-159.

Gower, J.F.R. and Borstad, G.A., 1981. Use of the in vivo fluorescence line at 685 nm for remote sensing surveys of surface chlorophyll a. In: J.F.R. Gower (Editor), Oceanography from Space, Marine Science, 13: 329-338. Plenum Press, New York.

Gower, J.F.R. and Lin, S., 1983. The information content of different optical spectral ranges for remote chlorophyll estimation in coastal waters, International Journal of Remote Sensing. In press.

Hovis, W.A., Clark, D.K., Anderson, F., Austin, R.W., Wilson, W.H., Baker, E.J., Ball, D., Gordon, H.R., Mueller, J.L., El-Sayed, S.Z., Sturm, B., Wrigley, R.C., and Yentsch, C.S., 1980. NIMBUS 7 Coastal Zone Color Scanner: System description and initial imagery. Science, 210: 60-63.

Hovis, W.A., 1981. The NIMBUS 7 Coastal Zone Color Scanner (CZCS) program. In: J.F.R. Gower (Editor). Oceanography from Space, Marine Science, 30: 213-225. Plenum Press, New York.

Kim, H.H., Hart, W.D. and van der Piepen, H., 1982. Initial analysis of OSTA-1 Ocean Color Experiment Imagery. Science, 218: 1027-1031.

LaViolette, P.E., Peteherych, S. and Gower, J.F.R., 1980. Boundary Layer Meteorology, 18: 159-175.

Lyzenka, D.R., 1981. Remote sensing of bottom reflectance and water attenuation parameters in shallow water using aircraft and Landsat data, International Journal of Remote Sensing, 2: 71-82.

Morel, A.Y. and Gordon, H.R., 1980. Report of the Working Group on Ocean Color. Boundary Layer Meteorology, 18: 343-355.

NASA, 1982. The Marine Resources Experiment Program (MAREX) Report of the Ocean Color Science Working Group. Goddard Flight Center. R. Kirk (Coordinator).

Neville, R.A. and Gower, J.F.R., 1977. Passive remote sensing of phytoplankton via chloropyll a fluorescence. Journal of Geophysical Research, 82: 3487-3493.

Richardson, P.L., 1981. Gulf Stream trajectories measured with free-drifting buoys. Journal of Physical Oceanography, 11: 999-1010.

Royer, T.C., 1981. Baroclinic Transport in the Gulf of Alaska Part I. Seasonal Variations of the Alaska Current. Journal of Marine Research, 39: 239-250.

SASC, 1983. NIMBUS 7 CZCS derived products scientific algorithm description. Report no. EAC-7-8085-0027. Systems and Sciences Corporation, Hyattsville, MD., USA.

SCOR, 1983. Remote Measurement of the Oceans from Satellites. Scientific Committee on Oceanic Research,.Working Group 70 report, in preparation.

Shannon, L.V., Mostert, S.A., Walters, N.M. and Anderson, F.P., 1983. Chlorophyll concentrations in the Southern Benguela current region as determined by satellite (Nimbus 7 Coastal Zone Color Scanner). Journal of Plankton Research, 5: 565-583.

Smith, R.C. and Wilson, W.H., 1981. Ship and satellite bio-optical research in the California Bight. In: J.F.R. Gower (Editor), Oceanography from Space, Marine Science, 13: 281-294. Plenum Press, New York.

Thomson, R.E., 1972. On the Alaskan Stream. Journal of Physical Oceanography, 2: 363-371.

Wright, C., 1981. Observations in the Alaskan Stream during 1980. NOAA Technical Memorandum ERL. PMEL-23.

CONTRIBUTION OF REMOTE SENSING TO MODELLING

Jacques C.J. NIHOUL
GHER, University of Liège, Belgium

1. Application of remote sensing to the identification of processes
   and structures and to the formulation of mathematical models of
   the marine system.

One of the most decisive contribution of remote sensing has
been the supplying, for the first time, of synoptic views of large
sea areas and the identification of mesoscale and macroscale horizontal structures which had been overlooked in field studies and
ignored in mathematical models.

Digital image analysis of Landsat data has revealed, for instance, the penetration in the Harima Sea (Japan) of a pair of
large scale vortices formed by amalgation of two series of coherent vortices, produced in the free boundary layers in the wakes
of the Naruto Straits'Capes. The vortex pair, apparently carried
along by the tidal currents in a first stage, was found to continue penetrating into the Harima Sea, after tide reversal, under
self-induced driving forces (Maruyasu et al., 1981).

This mechanism which plays a cogent role in local mixing could
not have been identified without synoptic remote sensing views of
the Seto Inland Sea.

NOAA 6 images of the Western Mediterranean have shown complicated seasonal circulation patterns, - including eddies, planetary
solitons, upwellings, fronts, water intrusions, coastal currents -
which could not have been apprehended by restricted experimental
investigations (e.g. Philippe and Harang, 1982, Preller and
Hulburt, 1982) (fig. 1).

The meandering of large scale currents like the Gulf Stream and
the subsequent shedding of synoptic eddies has never been properly
perceived and understood until remote sensing images of the area
were available (e.g. Behie and Cornillon, 1981).

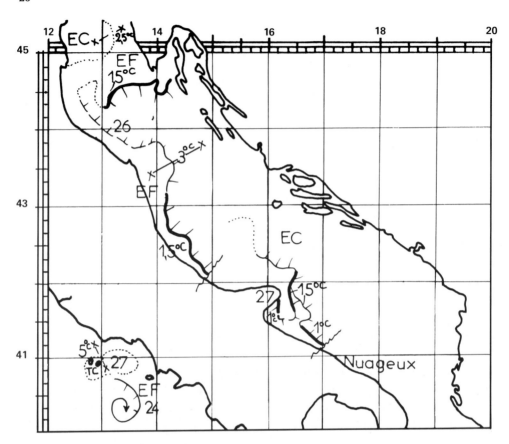

Fig. 1.   Chart of spring surface temperature fronts in the
Adriatic Sea (23-29 April, 1982) communicated by Lannion Center.

General legend for figures 1 to 4.

Mean position of a thermal front ($\Delta T \gtrsim 1\ °C$) persisting the
whole week; warm water on the dashed side.

Occasional thermal front ($\Delta T \gtrsim 1\ °C$) with the date of observation.

Permanent thermal border without marked frontal features.

Occasional thermal border without marked frontal features;
date of observation indicated.

Up   Upwelling

TC   Warm patch   DF   Cold divergence

EC   Warm water   EF   Cold water

Eddy   Nuageux   Cloudy

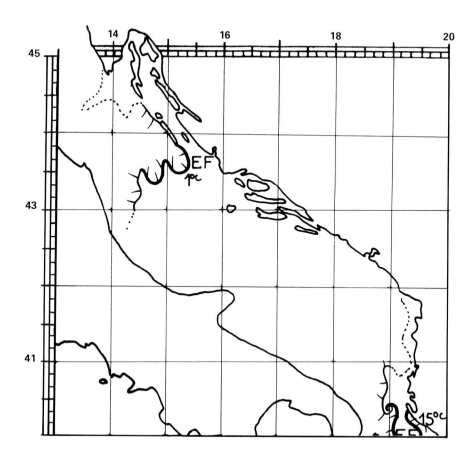

Fig. 2. Chart of summer surface temperature fronts in the Adriatic Sea (27 August - 2 September, 1982) communicated by Lannion Center.

The observation of extended patches of primary production, of turbid river plumes penetrating, meandering and spreading into coastal seas, - as well as the monitoring of marine pollution -, has come within reach with the development of remote sensing techniques (e.g. Ulbricht, 1981, Alberotanza and Zandonella, 1981, Horstmann and Hardtke, 1981).

There is no doubt that remote sensing has contributed to uncover synoptic processes and horizontal structures which limited field investigations could not apprehend and which mathematical models would have ignored.

The continuation of remote sensing surveys is essential to the development of reliable mathematical models based on a sound understanding of the dominant physical processes and an accurate formulation or parameterization of them.

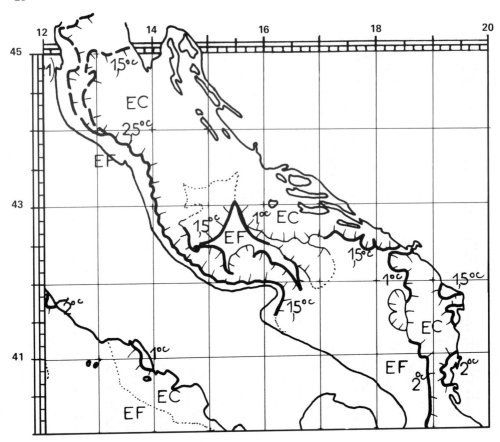

Fig.   3.   Chart of fall surface temperature fronts in the
Adriatic Sea (19-25 November, 1982) communicated by Lannion Center.

Two remarks should be made here :

1. The contribution of remote sensing to the identification of
cogent mechanisms which must be included in the formulation of
models, is more a question of a continuous and systematic surveying of the atmosphere and the ocean, - with sufficiently extensive
coverage and sufficiently detailed resolution -, than a question
of precise remote sensing measurements in specific regions.

Remote sensing is used here to unearth, elicit and identify,
for modelling, essential features of the marine system which,
because of time and length scales, are not easily detected by
classical field investigations. A routine rapid supply of all
available information is much more important than the delayed
delivery after careful image corrections of a limited set of remote sensing data.

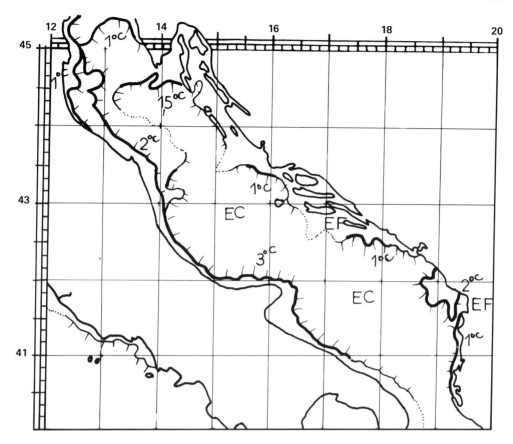

Fig. 4. Chart of winter surface temperature fronts in the
Adriatic Sea (18-24 February, 1983) communicated by Lannion Center.

2. The role of remote sensing in identifying processes for subsequent modelling increases with the marine system's inherent complexity (existence of frontal structures, eddies, ...). A useful
confirmation of the modeller's diagnostic, in the case of well-
mixed shallow tidal seas like the North Sea, the remote sensing
information becomes a vital prerequisit when one covets the modelling of deep, stratified seas like the Mediterranean with horizontal structures (gyres, fronts, eddies, ...) featuring the most
complex aspects of ocean hydrodynamics.

This must be kept in mind and given due priority when selecting
future test sites for case study applications of remote sensing
to mathematical modelling.

2. <u>Applications of remote sensing to the calibration, initiation</u>
   <u>and operation of mathematical models.</u>

A mathematical model is a set of partial differential equations
describing the evolution in space and time of the selected *state
variables*. These equations must be solved, subject to appropriate
initial and *boundary conditions*.

The limitation of the number of state variables and the formulation of the evolution equations and the boundary conditions
introduce several *parameters* which must be given, as semi-empiri-
cal functions or numerical values, before the model can be operated.

The determination of the parameters is based on theoretical reflections, sideways models and, decisively, on experimental data
derived from past or concurrent observations.

The boundary conditions which express interactions between the
system and the "outside world", inevitably depend on certain characteristics of the latter which appear as additional parameters.
This requires a supplement of experimental data.

Thus, experimental data are needed

(i)   to *calibrate* the model, i.e. to value the parameters, ta-
        king into account the distinctive regional features of the
        system ;

(ii)   to *initiate* the model, i.e. to specify the state variables
        at some initial time, when the simulation begins ;

(iii)   to *operate* the model, i.e. to determine the boundary condi-
        tions valid at all time of the simulation process.

Past experimental data, eventually accumulated over years, can
be used to calibrate a model. The model is then run, in a series
of *"hindcasting"* exercices, to simulate known situations of the
past and the parameters are adjusted to realize the best possible
agreement between the model's predictions and the observations.

Once a model has been calibrated, it can be used for forecasting provided initial and boundary conditions are given. One
emphasizes that the boundary conditions must be given *at all time*
of the simulation run. Thus, *model forecasting* is not equivalent
to *predicting the future* unless one can prognosticate the conditions at the boundaries. This happens when the boundary conditions are immutable (e.g. zero velocity at the coast) or when
they can be inferred, with sufficient precision, from the statistics of past similar situations.

TABLE 1

The GHER three-dimensional Mediterranean Sea model

State variables

Buoyancy $\quad b = - g \dfrac{\rho - \rho_0}{\rho_0} \quad (ms^{-2})$

Velocity $\quad v = u + v_3 e_3 \quad (ms^{-1})$

Pressure $\quad p \quad (kg\,m^{-1}s^{-2}) \quad$ or $\quad q = \dfrac{p}{\rho_0} + gx_3 + \hat{\xi} \quad (m^2 s^{-2})$

($\hat{\xi}$ is the tidal potential)

Turbulent kinetic energy $\qquad e \quad (m^2 s^{-2})$

Turbidity (or other passive or semi-passive constituent) $\qquad c = \dfrac{\rho_t}{\rho_0}$

Evolution equations

(1) $\quad \dfrac{\partial u}{\partial t} + \nabla.(uu) + \dfrac{\partial}{\partial x_3}(v_3 u) + f\,e_3 \wedge u = - \nabla q + \tilde{\mu}\,\nabla^2 u + \dfrac{\partial}{\partial x_3}(\tilde{v}\,\dfrac{\partial u}{\partial x_3})$

(2) $\quad \nabla.u + \dfrac{\partial v_3}{\partial x_3} = 0 \qquad ; \qquad$ (3) $\quad \dfrac{\partial q}{\partial x_3} = b$

(4) $\quad \dfrac{\partial b}{\partial t} + \nabla.(ub) + \dfrac{\partial}{\partial x_3}(v_3 b) = \tilde{\varkappa}^b\,\nabla^2 b + \dfrac{\partial}{\partial x_3}(\tilde{\lambda}^b\,\dfrac{\partial b}{\partial x_3})$

(5) $\quad \dfrac{\partial e}{\partial t} + \nabla.(ue) + \dfrac{\partial}{\partial x_3}(v_3 e) = \pi^e + \tilde{\varkappa}^e\,\nabla^2 e + \dfrac{\partial}{\partial x_3}(\tilde{\lambda}^e\,\dfrac{\partial e}{\partial x_3})$

$\qquad \pi^e = \tilde{v}\left\|\dfrac{\partial u}{\partial x_3}\right\|^2 - \tilde{\lambda}^b\,\dfrac{\partial b}{\partial x_3} - \alpha_\epsilon\,\dfrac{e^2}{16\,\tilde{v}}$

(6) $\quad \dfrac{\partial c}{\partial t} + \nabla.(uc) + \dfrac{\partial}{\partial x_3}(v_3 c) = \pi^c + \tilde{\varkappa}^c\,\nabla^2 c + \dfrac{\partial}{\partial x_3}(\tilde{\lambda}^c\,\dfrac{\partial c}{\partial x_3})$

$\qquad \pi^c = \begin{cases} 0 & \text{(passive constituent)} \\ - kc & \text{(decaying constituent)} \\ - \sigma\,\dfrac{\partial c}{\partial x_3} & \text{(sedimenting constituent)} \end{cases}$

$\qquad \nabla = e_1 \dfrac{\partial}{\partial x_1} + e_2 \dfrac{\partial}{\partial x_2}$

In principle, the value of the state variables within the system's support are calculated by the model and experimental data are only required there at the initial time. The more experimental data one has, however, the better and one must exploit all available information. Redundant data can be used to complete or up-date the model's calibration, compensate for missing boundary data or control the simulation.

TABLE 2

The GHER three-dimensional Mediterranean Sea model

Evolution parameters

Horizontal eddy viscosity : $\tilde{\mu}$ # (mesh size)$^{4/3}$

Vertical eddy viscosity : $\tilde{\nu} = 0.5\ \alpha_e^{1/4}\ e^{1/2}\ \ell$

Horizontal eddy diffusivity of buoyancy : $\tilde{\varkappa}^b = \alpha_b\ \tilde{\mu}$

Vertical eddy diffusivity of buoyancy : $\tilde{\lambda}^b = \phi^b\ \tilde{\nu}$

Horizontal eddy diffusivity of turbulent energy : $\tilde{\varkappa}^e = \alpha_e\ \tilde{\mu}$

Vertical eddy diffusivity of turbulent energy : $\tilde{\lambda}^e = \phi^e\ \tilde{\nu}$

Horizontal eddy diffusivity of turbidity : $\tilde{\varkappa}^c = \alpha_c\ \tilde{\mu}$

Vertical eddy diffusivity of turbidity : $\tilde{\lambda}^c = \phi^c\ \tilde{\nu}$

Constant of exponential decay : k

"Sedimentation" velocity : σ
(with the x_3 axis pointing upwards, σ is negative for sedimentation i.e. settling on the sea floor; σ is positive for rise of light constituents to the sea surface)

α_i is a dimensionless constant of order 1 , $i = e,b,c,\ldots$

f is the "Coriolis frequency" equal to twice the vertical component of the Earth's rotation vector. (f is a function of latitude; f may be approximated by a constant if the area is not too large; in mid-latitudes $f \sim 10^{-4}\ s^{-1}$).

$$\ell = H\ \xi(\tfrac{z}{H})\ \psi(Ri) \qquad ; \qquad Ri = \frac{\dfrac{\partial b}{\partial x_3}}{\left\|\dfrac{\partial u}{\partial x_3}\right\|^2} \equiv \frac{N^2}{M^2}$$

$$\phi^b = 1.4\ e^{-1.2\ Ri} \sim \phi^c \qquad ; \qquad \phi^e \sim 1 \qquad ; \qquad \psi = e^{-0.8\ Ri}$$

The empirical form of the function ξ is adjusted to comply with the logarithmic boundary layer requirements at the bottom and the wind mixed upper layer requirements at the surface [for a detailed discussion, see Nihoul (1982)].

The information required from remote sensing or sea-truth measurements depend on the scope and on the support of the system and their nature, scale and precision must be consistent with the model.

To identify the data requirements for the mathematical modelling of hydrodynamic and passive dispersion processes in a given marine and climatic environment, one must first install the proper mathematical model, bringing to light state variables and parameters.

TABLE 3

The GHER three-dimensional Mediterranean Sea model

Boundary_conditions_at_the_air-sea_interface

At $x_3 = \zeta$

$$\tilde{\nu} \frac{\partial \mathbf{u}}{\partial x_3} = C_0 (1 + 0.1 \|\mathbf{V}_{10}\|) \|\mathbf{V}_{10}\| \ \mathbf{V}_{10}$$

$$\tilde{\lambda}^b \frac{\partial b}{\partial x_3} = C_0 \ 10^{-3} \ \{\alpha_T \ \|\mathbf{V}_{10}\| \ (T_{10} - T_0) + \alpha_L \|\mathbf{V}_{10}\| \ \frac{L(c_{10}^v - c_0^v)}{c_p^a}$$

$$+ \frac{\alpha_s}{2} [1 - (1 - \gamma_s)m]F_0 - \frac{\alpha_l}{4} (1 - 5\sqrt{c_{10}^v})(1 - 0.6 \ m^2) \ \tilde{\sigma} \ T_0^4\}$$

$$+ \alpha_p g \ c_0^s \ F_p$$

$$\tilde{\lambda}^e \frac{\partial e}{\partial x_3} = 3 \alpha_E C_0 \ 10^{-3} \ \|\mathbf{V}_{10}\|^3 \quad ; \quad \tilde{\lambda}^c \frac{\partial c}{\partial x_3} = 0$$

Boundary_parameters_and_required_data

$\mathbf{V}_{10}$, T_{10} , c_{10}^v : wind velocity, air temperature, humidity at reference height

T_0 , c_0^s : sea surface temperature and salinity

F_0 : total short-wave irradiance of the sea surface in the absence of clouds

F_p : net precipitation flux (precipitation - evaporation)

m : cloud amount (in tenths)

γ_s : coefficient representing the effect of solar altitude on the amount of radiation that can get through a cloud cover (γ_s varies with the sun's altitude, its mean annual value is a function of latitude)

$\tilde{\sigma}$: Stefan-Boltzman constant ($\tilde{\sigma} \sim 5.67 \ 10^{-8}$ W m^{-2} K^{-4})

C_0 : a dimensionless coefficient $\sim 0.63 \ 10^{-6}$

α_i : a dimensionless coefficient of order one (i = T,L,s,...)

By way of illustration, the three-dimensional model developed by the Geohydrodynamics and Environment Research Laboratory (GHER) of Liège University to study hydrodynamic and dispersion processes in the Mediterranean and, in particular the Adriatic Sea (Nihoul, 1982) is summarized in tables 1, 2 and 3, with emphasis on the definition of state variables and parameters, the formulation of boundary conditions at the air-sea interface and the specification of the data requirements.

At first sight, the data situation, as far as remote sensing is concerned, appears rather disappointing.

Obviously some of the necessary data can be provided by remote sensing but many remain to be determined by conventional means. A single run of an airborne coastal zone colour scanner (CZCS), for instance, does not supply the necessary information on the wind stress, the air temperature and humidity at reference height ...   Would it be possible to retrieve such information from the mass of work which goes into the error correction process, the data would never be supplied in real time for model forecasting.

More sophisticated, present or future, satellites may provide part of the missing data but it is still not entirely clear whether (or when) the required accuracy and the minimum coverage needed for shelf sea studies will be achieved (e.g. Frassetto and Paci, 1981).

At this stage, it is fair to say that remote sensing images still appear more qualitative than quantitative to the modeller who tends to consider that their decoding does not provide data with sufficient accuracy.

One should however not misjudge the situation and the following remarks are appropriate :
1. Several remote sensing missions in the past had, for primary objective, the testing of a sensor or of a carrier. The interest lay essentially in decoding the signals, determining the appropriate atmospheric corrections and filing the observations as additional information on the environment. The experiments were not designed to provide the data required by mathematical modelling and, in many cases, the scientific product of such remote sensing experiments - while definitely contributing to complete the data base available for the calibration of the model - is not appropriate to initiate and operate mathematical models.
2. A second generation of satellites and sensors has demonstrated the accessibility of more complete data sets for modelling applications.

A combination of five instruments, a radar altimeter (ALT), a microwave scatterometer (SASS), a synthetic aperture radar (SAR), a visible and infrared radiometer (VIRR) and a scanning multi-channel microwave radiometer (SMMR) enabled SEASAT to supply useful data for modelling and to provide valuable information on ocean-modelling-oriented future sensors (e.g. Born et al., 1981). "Space Oceanography" is now entering a more mature phase with better, more diversified, equipment which takes into account the requirements of modelling.

This has been made possible by the pioneer effort to use the available, - often exploratory -, remote sensing data to provision modelling endeavours. It is imperative that this effort be continued when it begins to bear fruits.

3. Because decoding and interpretation of the results of any particular sensor is such a formidable task, one has often forgotten that modelling was a constant applicant for all (remote sensing) available data and models of certain marine regions have been undertaken without the necessary preliminary collection and processing of all pertinent remote sensing information.

One may expect that, in the next phase of routine survey, the accessibility of data from all sources and a world-wide distribution system getting into its stride will permanently purvey an international network of marine, atmospheric and climatic models, forecasting from remote sensing continuous information.

In the same time, satellite remote sensing surveys will be more and more complemented by specific purpose airborne missions and permanent land-based radar monitoring of winds, waves and surface currents associated with guided sea-truth measurements.

One may thus conclude that the contribution of remote sensing to the calibration, the initiation and the operation of mathematical models, - although not yet self-sufficing -, is valuable and continuously improving. Models will rely more and more on specially designed remote sensing investigations and the endeavours of to-day - however limited by unadequate remote sensing data, in this preliminary testing phase -, are paving the way for to-morrow's concerted development of remote sensing and modelling.

REFERENCES

Alberotanza, L., Zandonella, A., 1981. Landsat imagery of the Venetian Lagoon . A multitemporal analysis. In : J.F.R. Gower (Editor), Oceanography from Space. Plenum Press, New York, pp. 421-428.

Behie, G., Cornillon, P., 1981. Remote sensing, a tool for managing the marine environment. Eight case studies, University of Rhode Island. Marine Technical Report 77, 44 pp.

Born, G.H., Lame , D.B., Rygh, P.J., 1981. A survey of the goals and accomplishments of the seasat mission. In : J.F.R. Gower (Editor), Oceanography from Space. Plenum Press, New York, pp. 3-14.

Frassetto, R., Paci, R., 1981. The proposed coastal monitoring satellite system of ESA : mission objectives and problems. In : J.F.R. Gower (Editor), Oceanography from Space. Plenum Press, New York, pp. 29-34.

Horstmann, U., Hardtke, P.G., 1981. Transport processes of suspended matter, included phytoplankton, studied from Landsat images of the Southwestern Baltic Sea. In : J.F.R. Gower (Editor), Oceanography from Space. Plenum Press, New York, pp. 429-438.

Maruyasu, T., Onishi, S., Nishimura, T., 1981. Study of tidal
    vortices at the Naruto Strait through remote sensing. Bulletin
    of the Remote Sensing Laboratory, Remote Sensing Series n° 1.
    The Science University of Tokyo Publ., Tokyo, 142 pp.
Nihoul, J.C.J., 1982. Data requirements from remote sensing and
    sea-truth for hydrodynamic passive dispersion models. Joint
    Research Center, ISPRA, R 1880-82-06 ED ISPB/GNO4, 20 pp.
Philippe, M., Harang, L., 1982. Surface temperature fronts in
    the Mediterranean Sea from infrared satellite imagery. In :
    J.C.J. Nihoul (Editor), Hydrodynamics of semi-enclosed seas.
    Elsevier, Amsterdam, pp. 91-128.
Preller, R., Hurlburt, H.E., 1982. A reduced gravity numerical
    model of circulation in the Alboran Sea. In : J.C.J. Nihoul
    (Editor), Hydrodynamics of semi-enclosed seas. Elsevier,
    Amsterdam, pp. 75-89.
Ulbricht, K.A., 1981. Examples of applications of digital image
    processing of remotely sensed phenomena. In : A.P. Cracknell
    (Editor), Remote Sensing in Meteorology, Oceanography and
    Hydrology. Ellis Horwood Publ., Chichester, pp. 97-107.

OPTIMAL REMOTE SENSING OF MARINE ENVIRONMENT

I.V. MURALIKRISHNA

Marine Applications Division, National Remote Sensing Agency
Hyderabad : 500037 (India)

ABSTRACT

The comprehensive data base required to understand the marine environment exceeds the information which could be acquired from satellite remote sensors. One can optimize a remote sensing campaign by planning the remote sensing measurements from space in conjunction with the ground based components. Such an optimal approach was used in a study of littoral processes along the east coast of India. On similar grounds, extensive observations of sea-surface temperature for validation of thermal infrared measurements from space are suggested. Also, the satellite data correspond to surface manifestations and it is appropriate to identify the role of in situ data in the context of satellite data availability. The combination of in situ and satellite data can possibly lead to an assessment of the three dimensional configuration of the marine environment. This forms the basis for defining what is called optimal remote sensing.

INTRODUCTION

Ten years ago, that is in the early 70s, particularly after the launch of the Landsat satellite, scientists all over the world thought that remote sensing would provide a solution to all the problems concerned with the evaluation of natural resources and the monitoring of the environment. But two years later, they came to the conclusion that remote sensing is not as useful as envisaged and that it is hampered by numerous problems. The main reasons are the isolation of the space observing systems form the ground based components and the lack of adequate transfer functions that would lead to the evaluation of the resources from the otherwise irrelevant series of digital data recorded according to a particular format on a computer compatible tape. This forms the essential background philosophy in defining what is called optimal remote sensing. It is more or less a concept. The scientific community has now realised that remote sensing can provide large datasets and be very useful when used in conjunction with field surveys and in situ monitoring of a few test sites. The foregoing is particularly

applicable for remote sensing of any dynamic feature which needs
analysis in real time. This is what is accepted as optimal remote
sensing. At present, there is no law prohibiting remote sensing,
but remote sensing has not received a specific legal clearance,
either.

While it is a fact that space-based observations are ideally
suited for such oceanographic purposes as evaluating the directional wave energy spectrum or the horizontal distribution of chlorophyll, remote measurements generally bring out only incomplete and
indirect answers to such complicated questions as the assessment of
fisheries yield or the understanding of wave climate. Hence the
effectiveness of the supplementary space observations cannot be
gauged directly against established data requirements. One can
optimized a remote sensing operation by planning the measurements
from space in conjunction with other ground based components
rather than in isolation from the continuing refinements of data
requirements normally needed for advanced research in marine
science. In other words, remote sensors operating from the vantage
point of space will never replace direct measurements, because the
ocean is more or less opaque to electromagnetic radiation. However,
satellite remote sensing, data relay, and platform location
techniques should play a significant role that needs to be systema-
tically recognized and exploited in future programs of marine
sciences research. Recent experience with sensors on GOES-3, Seasat
and Nimbus-7 designed for ocean observations, underlines the need
to include from the beginning, explicit planning for validation/
control observations, and a substantial data collection effort. To
do otherwise would be to take a chance of not extracting the full
advantage of the very large investment in the satellite portion of
the system (Ruttenberg, 1981).

RELEVANCE OF MARINE DATA AND INFORMATION SYSTEM

The comprehensive data base required to understand the marine
environment exceeds the information which could be acquired from
satellite remote sensors. This difference is expected to decrease
with the development of active microwave remote sensors together
with satellites to carry those sensors and algorithms to transfer
the remotely sensed data into meaningful oceanographic parameters.
One of the major problems of marine resource data is the necessity
of taking fully into account the real time nature of the data flow
and processing. This is particularly true of the data connected

with marine bio-chemical and bio-optical properties. That is, any oceanographic information must be thought of as resulting from a real time processing effort, rather than in terms of storage for subsequent leisurely consideration. Conventionally, the oceanographers have been oriented towards vertical sampling, and horizontal structures used to be studied based on data from a few sampling stations. To the contrary, remote sensing techniques should possibly provide information that has basically new characteristics, such as horizontal averaging over larger regions and the feasibility of averaging over repeated observations. The limitations associated with inference from remote sensing, and the difficulties of reconstructing the overall picture from limited in situ observations imply that the acceptance of the new information will come only after systematic studies involving remote sensing and in situ data collection.

Application of remote sensing to the study of the "bio-optical state" of the sea

There is an urgent need to include, from the beginning, the planning of specific data collection efforts for validation, calibration and analysis of the satellite data. The satellite remote sensing can change the hitherto prevailing concepts of marine productivity. However this is only possible after the necessary corrections have been applied and calibrations and error estimates are known. For example, in marine ecology, the main problem is to establish both spatial and temporal scales in which physical and biological processes occur.

The biological processes have significant effect on the optical properties of seawater. Hence a state called "bio-optical state" is the significant parameter that requires detailed analysis. Taking chlorophyll concentration as an index of primary productivity, remote sensing from aircraft and satellites provide exclusive oceanographic information. Data collected from ships provide more accurate information, but such information is limited in space and time scales. Aircraft and satellites, on the other hand, can provide synoptic information whose accuracy is the subject of ongoing research and likely to improve. A comparison of various pigment evaluation algorithms suggests that our ability to relate pigment concentration to optical characteristics is seriously impaired by the relatively small quantity of good quality optical data on which any model can be tested (Muralikrishna, 1983). Also,

the proposed models work well only for those specific waters for
which the algorithms were developed. This element of uncertainity
will prevail until the models take into consideration all possible
parameters and untill exhaustive sea truth programs are carried
out.

According to Yentsch (1983), for anyone concerned with coastal
and ocean processes the value of satellite observations cannot be
overestimated. Based on a few conventional vertical profiles of
chlorophyll, salinity, temperature and other biochemical parameters, oceanographers are able to construct a total three dimensional picture of the ocean and assess the productivity and bio-optical state. This is important because vertical mixing is crucial
to phytoplankton growth and distribution (Yentsch, 1983). Now that
remote sensing satellites provide at specified time intervals information such as horizontal distribution of concentration and its
gradients over larger areas, it should be possible to construct a
three dimensional picture of the ocean as done by conventional
survey provided a trade-off between horizontal structure and vertical sampling is established. This is not possible on a small
scale study. Such an effort requires simultaneous in situ collection of vertical profiles of several bio-chemical, optical and
physical parameters and satellite sensor data. The satellite sensor should have channels narrower than those of the Landsat multispectral scanner with a channel in the blue part of the spectrum
and a spatial resolution of the order of 30 to 40 meters. This
study should yield information regarding how much in situ data is
required to assess the total vertical structure of some oceanic
region from satellite data. The questions to be answered are :
(1) is this information dependent on the type of biochemical parameter under study; (2) is it season dependent, or (3), location
dependent ? This is a very important step in linking conventional
oceanographic studies with remote sensing studies and in establishing
the validity of the latter. Such a study is only possible through
multi-organisational (bcth national and international) cooperation
under the sponsorship of agencies like the Intergovernmental
Oceanographic Commission. This is a suggestion for possible future
cooperation.

Another aspect to consider is marine pollution. First of all
some reference data base involving satellite, aircraft and in
situ studies should be created for all areas that are prone to
environmental degradation. For example, consider the case of oil

slicks; it is essential to conduct an optical, and microwave remote sensing survey to establish the potential of remote sensing to monitor oil pollution.

Application of remote sensing to the study of physical oceanographic and engineering parameters

It is evident that there is a need for further research in this area to establish the characteristics of physical oceanographic and engineering parameters as seen by a satellite based sensor. The temperature of the sea surface is one of the important physical factors that determine the exchange of heat energy between the atmosphere and the ocean. A reliable and coherent sea-surface temperature data set is badly needed. Satellite observations are the only feasible way to achieve this. The present NOAA-7 and NOAA-8 satellites offer, in principle, the promise of providing this information, provided that data processing can be improved well beyond the present system. In order to validate the data evaluation method and the results, it is essential to measure sea surface temperature at a minimum of about 40 to 50 locations simultaneously. To obtain simultaneous observations of temperature values, say in the Arabian sea, from NOAA-7 or NOAA-8 AVHRR, the processed data should yield the sea surface temperature to an accuracy of 1°C. A well coordinated in situ study by merchant and research ships at about 40 to 50 stations would be very useful for the validation of the satellite data processing method and of the results. For fisheries application, on the other hand, a wide spatial coverage and gradient information would be enough, the absolute value of the temperature being less important. During the past decade, microwave radiometry from space has developed into a powerful technique for remote sensing of the earth's atmosphere and of the oceans. A review documenting the significant observations of sea surface temperature discusses the scope of future applications (Njoku, 1982). Some of the most significant parameters that require to be regularly monitored for oceanic modelling and surveying purposes are :

1. wind stress;
2. sea level changes;
3. wave height;
4. directional wave spectrum;
5. heat budget of the ocean-atmosphere system.

The data provided by the Seasat based scatterometer may be of use to measure the wind stress. However, many special studies will be needed to improve the interpretation of the scatterometer obser- vations, particularly due to the dependence of the signal on a variety of phenomena like short and long wave interactions and short crested waves. The radar altimeters on the GOES-3 and Seasat satellites have demonstrated the ability to provide significant wave height. The synthetic aperture radar (SAR) images from Seasat also appear capable of providing important information on many oceanic features. However, it became evident that SAR does not always image similar surface waves in the same manner. Qualitative- ly,the relations derived to detect ocean waves in Seasat SAR images are consistent as summarised by Kasischke (1980). A more quantitative test .of the analysis, however, requires simultaneous measurements of sea truth and the backscatter modulation transfer functions which determine the modulation depth of the SAR wave image (Alpers, Ross and Rufenach, 1981). This is again within the requirements of optimal remote sensing as defined earlier in this paper. It is expected that the required information will become available through the set of SAR flights and sea truth measurements conducted during the Marine Remote Sensing Experiment (MARSEN). Some specific results of these studies are evaluated by Alpers and Hasselman (1982). These studies highlight the need for the develop- ment of processing techniques and the systematic development of transfer functions to obtain ocean parameters from SAR data. This is to be taken as the requirement of oceanographers to use remote sensing. In view of the proven utility of Seasat SAR, altimeter, and scatterometer, these active microwave remote sensors should be recommended for future ocean monitoring satellites. The only satel- lite with this type of payload on the anvil as of today is ERS-1 (European remote sensing satellite). It would be worthwhile to plan a well-coordinated ship based in situ data collection program that can be used as a sea truth survey for ERS-1. This would help to develop the necessary algorithms and to establish the validity of the results by comparison with field data.

Following this review of the potential and significance of satellite remote sensing observations, we consider a case of appli- cation of satellite data to monitor the marine environment. The multispectral scanner (MSS) onboard the Landsat satellite has 4 bands in the visible and near-infrared region of the electromagne- tic spectrum. A set of scenes from Landsat has been studied

together with in situ optical and hydrographical data collected
near Paradip, Orissa, off the east coast of India. The main aim is
to assess the effect of littoral processes and to estimate the
accretion and erosion rates over a given period of time.

LITTORAL PROCESSES

The transport of sediments in the littoral zone by waves and currents is called littoral transport. The littoral zone is the zone
extending from the shoreline to just beyond the nearshore zone.
Littoral transport is classified as onshore-offshore transport or
as longshore transport. In the nearshore region the onshore-offshore
transport is predominant and in the survey region both longshore
and onshore-offshore transports are significant. Engineering applications involving littoral transport generally require solutions
to problems regarding the conditions of longshore transport at the
site and shoreline migration. The measurement and analysis of
combined beach and nearshore profiles is a major part of most engineering studies of littoral processes. In combination with beach
profiles, repetitive neashore profiles are used in coastal engineering to estimate erosion and accretion along the shore. Under
favourable conditions, nearshore profiles have been used to measure
longshore transport rates. Making use of this practice, a study
has been carried out to assess the capability of remotely sensed
satellite data for monitoring the effects of littoral processes.

Landsat MSS data are used for this study in conjunction with in
situ optical data and navigation charts. The study area is located
near Paradip, along the east coast of India. The methodology adop-
ted to monitor littoral processes is based on the comparison of
nearshore profiles. The satellite data are expected to give nearshore profiles to an accuracy limited by spatial and spectral
resolution and concentration of suspended sediments. One of the
objectives of this study is also to explore the potential of MSS
data to evaluate nearshore processes and to identify the possible
effects of existing structures. A sample Landsat image is shown in
Fig. 1. This is an MSS 7 image (infrared) which is suitable for
the delineation of land/water boundary. MSS 4 and 5 give good
penetration and are useful for the evaluation of bottom contours
in the nearshore region. The area under study is indicated by a
white square line in Fig. 1. The MSS data in the form of CCT for
this region has been "density sliced". The corresponding in situ
optical data was correlated with hydrographic data and a wedge in

44

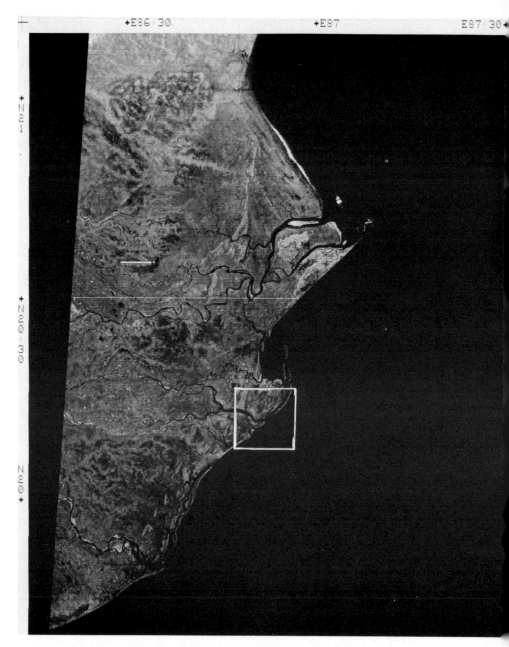

Fig. 1. Landsat MSS-7 scene no. 149-046 of 4 nov. 1981

Fig. 2. Density sliced image for part of scene no. 149-046

the form of various density levels was prepared. The density sliced image for the area marked in Fig. 1 is shown in Fig. 2. Since this corresponds to infrared wavelengths (MSS-7),only a few levels can be discerned. This figure,on a scale of about 1 to 25,000, is quite suitable for coastline comparison studies. The density sliced technique for MSS-4 and MSS-5 gave depth contours ranging from 0 to 12 meters.

A comparison of the depth contours near Paradip harbour indicated an advance of the deep water contour on the southern side of the harbour. During the August 1981 cyclones, the Mahanadi River cut across the coast line in the area denoted A in Fig. 2, and reached the sea north of the harbour. The Paradip harbour is shown as zone B in Fig. 2. The sea wall on the northern side of the harbour stands as a barrier against direct impact of wave attack and helps the shoreline to be fairly stable even during rough weather and cyclonic periods. The satellite data for the period 1975 and 1981 indicate that there is considerable erosion in the region north of the sea wall. For the region on the southern side of the harbour, the comparison of satellite data for the period 1975 to 1981 shows that the 5 and 10 meters contours moved seawards. This could be due to the obstruction of the normal drift by the southern breakwater.

Any coastline can maintain its equilibrium profile so long as there is no intrusive structure built in the vicinity of the beach. It would be essential to determine the annual quantity of drift and the net rate of erosion along the coast caused by the obstruction of this drift. The effect of the sea wall on the northern side is that the profile is fairly stable for most of the region except for a small seaward displacement of the 5 to 10 meters depth zone. This could possibly be due to the drift of sediment material discharged by the river. Since this is a fairly deep water region, the waves may not have sufficient energy to carry the material in the northward direction. This causes erosion in the region beyond the sea wall. On the basis of available information, it is evident that the Mahanadi River has a tendency to shift its mouth from south to north due to intense littoral drift along the coast, resulting in the formation of a long and narrow sandspit along the coast. When the river follows a long course parallel to the coast along this sandspit before reaching the sea, there is a tendency for the river to breach the sandspit and to cut short its course to the sea during periods of high seas and during the monsoons. During

the August 1981 cyclone, the Mahanadi River cut across the sandspit at a distance of 6 km form the harbour. The width of the cut is
about 1.2 km. The effect of this phenomenon on the maintenance of
the harbour entrance channel needs specific study. The accretion
near the southern breakwater for a coastline of 1 km length is
about 150,000 cubic meters. This value indicates only the order of
magnitude of the effect of the southern breakwater in arresting the
transport of littoral material by natural processes. These studies
are discussed in detail by Muralikrishna et al. (1983).

These studies demonstrate the utility and the limitations of the
Landsat MSS to monitor littoral processes. During approximately a
decade of Landsat applications, the development of analysis algorithms and monitoring has been a very active area of research.
These studies were possible despite the fact that the Landsat MSS
was designed for mapping land resources, and was not optimised for
remote sensing of ocean parameters. The improved spatial resolution
and spectral allocations of the sensors on the forthcoming series
of satellites to be launched during this decade should be able to
provide more useful real time data for monitoring dynamic phenomena such as coastal processes.

SOME SPECIFIC PROBLEMS TO BE ADDRESSED FOR MARINE REMOTE SENSING

As already indicated, optimality implies direct measurements
plus remote sensing from space. Some specific problems that need to
be addressed are :

i) Extensive observations of sea surface temperature sould be
carried out for the validation of thermal infrared and passive
microwave measurements from space.

ii) Of late, satellites have provided the first source of synoptic data on oceanic processes in the shelf and slope regions.
Because such data are available only for short periods, the annual
and seasonal differences have not yet been assessed. Thus the
information derived from satellites is to be used in close coordi-
nation with direct observations and the possible use of historic
observations for the evaluation of oceanic processes is to be
explored.

iii) A well coordinated in situ data collection study is needed on
an international level so as to serve as sea truth for ERS, as
JASIN (Joint Airsea interaction project) served as sea truth for
Seasat.

iv) Satellite data correspond to surface manifestations and it is necessary to devise a program in which the role of in situ data collection has to be assessed in the context of satellite data availability. This is mainly to specify the minimum extent of in situ data required to obtain a three dimensional picture of the ocean from remotely sensed data.

v) Development of exclusive ocean remote sensing satellite systems should take place with : (a) optical sensors having channel allocation in the blue part of the spectrum, narrower channels and finer spatial resolutions, (b) microwave sensors like SAR, Scatterometer and passive radiometer.

ACKNOWLEDGEMENTS

Thanks are due to Prof. B.L. Deekshatulu, Director, NRSA for his encouragment and for giving permission to present these studies. The guidance given by Dr. S.Z. Qasim, Secretary, Department of Ocean Development and Dr. Selim Morcos, Division of Marine Sciences of UNESCO is gratefully acknowledged. Thanks are due to Prof. C.S. Yentsch of the Bigelow Laboratory for Ocean Sciences, Maine, USA and Dr. Jim F.R. Gower, Institute of Ocean Sciences, Sidney, British Columbia, Canada for their encouragement. The author thanks specially Prof. J.C.J. Nihoul, University of Liège, Liège, Belgium, for showing keen interest in the author's work and for providing the necessary financial support. Thanks are also due to Dr. L.R.A. Narayan, Director (Applications), NRSA, for his valuable suggestions, and to Miss Sarala Dorothy for typing the manuscript.

REFERENCES

Alpers W. and Hasselmann K., 1982. Signal to clutter and noise properties of ocean wave imaging SARs. International Journal of Remote Sensing, Vol. 3, N° 4, p. 423.
Alpers W.R., Ross D.B., and Rufenach C.L., 1981. On the detectability of ocean surface waves by real aperture radar. J. Geophysical Research, Vol. 86, p. 6481.
Kasischke, E.S., 1980. Extraction of gravity wave information from space borne synthetic aperture radar data. M.Sc. Thesis, University of Michigan, Ann Arbor, Michigan.
Muralikrishna, I.V., 1983. Ocean color studies in Arabian Sea. In Remote sensing applications to marine science and technology, Chapter 14, ed. A.P. Cracknell, NATO-ASI., Series.
Muralikrishna, I.V., Laxminarayana, M., Mohana Rao, M., Meenakshisundaram, V. and Rao, G.S., 1983. Monitoring effect of littoral processes using remote sensing techniques. Proceeding of National Natural Resources Management System Seminar, Hyderabad, India.

Njoku, E.J., 1982. Passive microwave remote sensing of the earth
   from space - A Review. Proceedings of IEEE, Vol. 70, n°7,
   July, p. 728.
Rutterberg, S., 1981. Needs, opportunities, and strategies for a
   long term oceanic sciences satellite program. Report to NASS by
   NOSS Science Working Group, Boulder, Colorado, 72 pp.
Yentsch, C.S., 1983. Remote sensing of biological substances. In
   Remote Sensing applications to marine science and technology,
   Chapter 13, ed. A.P. Cracknell, NATO-A.S.I., - Series D. Reidel
   Publishing Company.

SATELLITE AND FIELD OBSERVATIONS OF CURRENTS ON THE EASTERN
SICILIAN SHELF

E.BÖHM[1] and E.SALUSTI[2]

[1]Dipartimento di Fisica, Università "La Sapienza", Roma (Italy)

[2]INFN, Dipartimento di Fisica, Università "La Sapienza", Roma
(Italy)

ABSTRACT
    This note describes a cold patch observable by thermal satellite
imagery in the Strait of Messina and along the eastern Sicilian
shelf. The phenomenon is particularly marked and perists all year
long, excluding winter. This patch has been investigated by means
of both thermal imagery and field observations. As a result, the
region near the Sill of the Strait of Messina has been found to
consist of peculiar marine water generated by the tidal mixing
of surface and deep layers of Atlantic water with Levantine water.
This marine water has a flux of 0.1 Sv and flows southward for
~ 100 Km along the Sicilian shelf.

1-INTRODUCTION
    This note describes a strong "upwelling" observable by satellite
imagery in the Strait of Messina and in the eastern Sicilian shelf
(fig.1). This is a particularly strong phenomenon: Philippe and
Harang's classical study (1982) shows that the cold patch remains
intense and persistent throughout the year, excluding wintertime.

    The patch was first investigated by thermal imagery and then
by means of field observations during the cruise PRIME '82 (§ 3).
It was found that there is a peculiar marine water (called "C
water" in the following) in the Strait of Messina which is generated by tidal mixing of surface (mixed layer) and deep (up to ~ 150m)
Atlantic waters with Levantine water. This water then flows for
~ 100 Km southward on the eastern Sicilian shelf. Its flux can be
estimated as ~ 0.1 x $10^6 m^3/s$ (Böhm et al., 1983). Theoretical
and experimental investigation of this tidal generation mechanism
has revealed that the interface between Atlantic and Levantine
water has vertical displacements of more than 100 m, with semi-

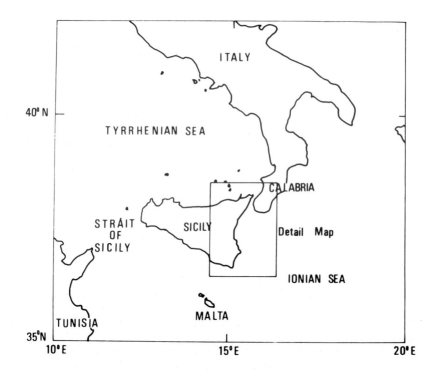

Fig.1a . Geographical location of the Strait of Messina.

diurnal tidal periodicity (§ 4).

This region has been fairly well known since F.Vercelli carried out a very careful set of measurements on the tidal currents in the geographical regions around the sill in 1922-23. Large baro-tropic/baroclinic motions have been observed and tidal mixing and related currents are known to occurr (§ 5).

2. SATELLITE THERMAL IMAGERY

Satellite thermal images of the Strait of Messina and the eastern Sicilian coastal areas are described below. The images were obtained by TIROS-N and NOAA 6 and processed digitally at the Ecole des Mines facilities (Sophia Antipolis, France). Their characteristics are shown in Table 1.

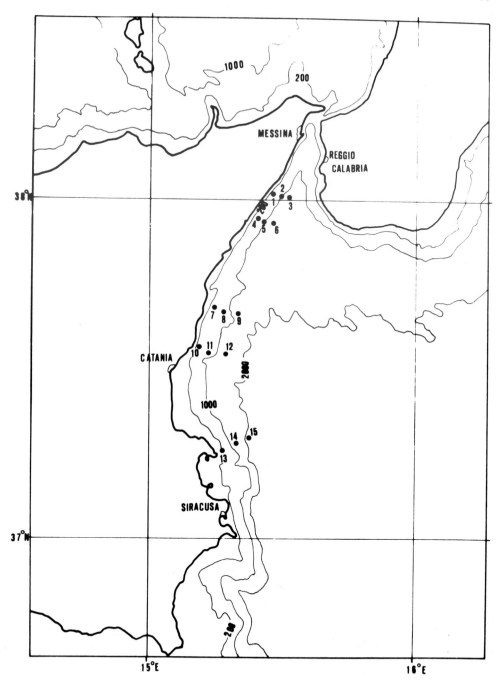

Fig.1b . Bathymetry and hydrographic stations of the eastern
Sicilian shelf and of the Strait of Messina (depths in meters).

TABLE 1
Characteristics of the 10 satellite images , divided into two tidal courses (northwards and southwards)
Tidal phase is computed according to Del Ricco (1982) treatment.

Satellite	Day	Time (GMT)	Tidal phase	Estimate of ΔT	COLD SPOT south border	COLD SPOT north border	COLD STRIP lenght	COLD STRIP width
				NORTH FLOWING CURRENT				
N6	7-4-81	6:59	8	(1.0 ± 0.1) C	23Km	8Km	90Km	4-10Km
N6	7-4-81	18:18	6	(0.8 ± 0.1) C	25Km	10Km	100Km	4-10Km
N6	7-7-81	7:36	5	(1.2 ± 0.1) C	22Km	0Km	missing	missing
TN	9-8-80	14:27	6	(2.2 ± 0.1) C	23Km	5Km	missing	missing
TN	10-8-80	14:14	7	(2.0 ± 0.1) C	20Km	1Km	missing	missing
N6	18-9-80	6:40	3	(0.6 ± 0.1) C	missing	missing	50Km	2- 4Km
				SOUTH FLOWING CURRENT				
N6	19-5-81	7:49	10	(1.4 ± 0.1) C	15Km	0Km	65Km	4-10Km
N6	2-6-81	7:32	10	(2.2 ± 0.1) C	85Km	5Km	85Km	4-10Km
N6	14-7-80	7:37	9	(3.6 ± 0.1) C	14Km	0Km	65Km	vortex
N6	16-7-81	7:32	11	(1.2 ± 0.1) C	15Km	3Km	20Km	5Km

According to Philippe and Harang (1982) no significant sea surface temperature difference is detectable in wintertime , in this imagery. Two different phenomena are visible in the spring, summer and fall images. The first of these is a cold-water spot bounded by a thermal front ($\Delta T \sim 2\,°C$) ranging from a few Km north to 15 - 25 km south of the sill (Fig.2). The second phenomenon is a cold coastal water strip extending southward from the sill (up to 100Km long and 4 to 10 Km wide) following the sharp shelf of Sicily (Fig.3). These two phenomena are often observed simultaneously.

Analysis of these images has shown that the cold-water spot is linked with the tide which mixes Atlantic water masses with Levantine water masses.

The cold strip is not always completely visible in the satellite images but, when it does appear in its full extension, it can show meanders and eddies (Fig.4). From the displacement of the southernmost location of the thermal front bounding the strip observed in a couple of successive images (eleven hours apart, April 7, 1980, 0659 GMT and 1808 GMT) the velocity normal to the front was estimated to be 0.2 m/s. It is interesting to note that, in this case, the C water was detected off Augusta, i.e. ~ 100 km away from where it was generated.

3-HYDROLOGICAL DATA

Hydrological data were collected at 15 stations along the sharp shelf of the eastern Sicilian coast on May 30 and 31, 1982, during the oceanographic cruise PRIME by the R/V Bannock of CNR (Fig.1b). One of the aims of the cruise was to investigate the strip of C water already observed in satellite thermal images. In each station, salinity, temperature, density and depth were measured by means of a Neil Brown bathysonde. The ten-meter depth temperature data were used to plot isotherms (Fig.5) similar to the thermographies already shown. A north-south thermal gradient can also be seen (previously observed by Grancini and Magazzù, 1973).

The T-S diagrams show the presence of two different waters - the C water corresponding to the cold-water strip closer to the Sicilian

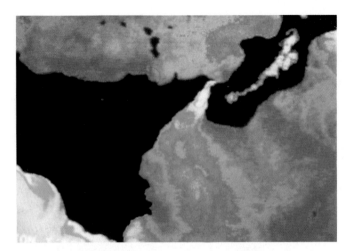

Fig. 2.
TIROS N image of a cold patch south of the Strait of Messina (Aug. 9, 1980; 1427 GMT).

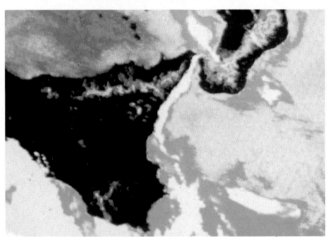

Fig. 3.
NOAA 6 image of a cold strip on the eastern Sicilian shelf (June 2, 1981; 0732 GMT).

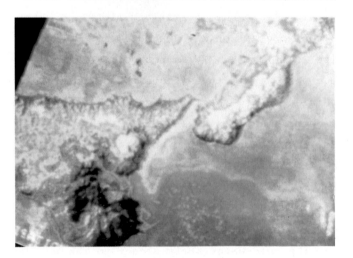

Fig. 4.
NOAA 6 image of a meander of the coastal current (April 7, 1981; 0659 GMT).

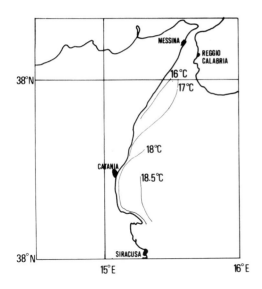

Fig. 5. Isotherms at a depth of 10 m , obtained from the hydrological stations of the PRIME cruise.

coast and the Ionian water offshore. The T-S diagrams have been
divided into three types (Fig.6a):

   a) C water only (stations 1, 5 and 4);

   b) C water covered by mixed C and Ionian waters (stations 2,6,
      7, 8, 10, 11, 13 and 14);

   c) Ionian water only (offshore stations 3, 6, 9, 12 and 15).

   The main difference between C and Ionian water is that the
former is stratified both in temperature and salinity, while the
latter is stratified only in temperature. The offshore border
between Ionian and C waters was found to be 5 - 10 km offshore.
Hydrologically speaking, C water ranged from t=16.0 °C and S=38.2 ‰
to T=14.0 °C and S=38.7‰ over a thickness of approximately 100 m.

   The ship's drift data taken under windfree conditions gave a
prevailing southern orientation of currents that was weakly dependent on the across-shore distance. The absolute velocity values
ranged from 0.5 m/s (in the northernmost location) to 0.2 m/s.
These data allow a rough but nevertheless interesting estimate to
be made of the southward flux of C water, $\Phi = (1.5 \pm 0.3) \cdot 10^5 \, m^3/s$

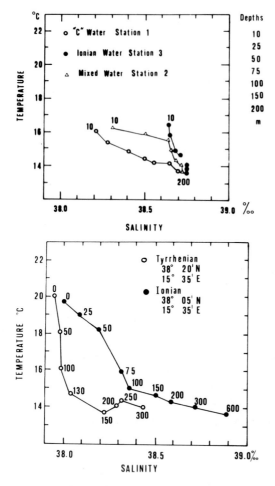

Fig.6a . T-S diagrams of types a), b), c).

Fig.6b . The vertical T-S structure from two hydrographic stations taken by Vercelli in Tyrrhenian and Ionian Seas.

in the northernmost section.

A detailed analysis of our T-S diagrams reveals two interesting features. First there is a remarkable variability above the seasonal thermocline due to the complex "history" of the C water. Moreover, below the seasonal thermocline, the T-S diagrams of the C water show a surprising similarity with the T-S diagrams of the Tyrrhenian surface water. This similarity suggests that, for the above depth range, the main component of C water is Tyrrhenian water.

4-TIDAL CURRENTS AND MIXING IN THE STRAIT OF MESSINA

Field measurements have shown that the cold strip observed by satellite thermal imagery is not an upwelling but a cold-surface

current flowing southward from the region around the sill of the
Strait of Messina. This suggests that the current originates in the
regions around the sill as a result of tidal mixing. Let us now
examine more fully the state of our knowledge of the tidal currents
and mixing in the Strait of Messina.

The cross section of the Strait of Messina is smallest over the
sill, between Punta Pezzo and Ganzirri. The cross-sectional area is
0.3 km^2, with a mean depth of about 80 m and a sill depth of about
120 m. From the sill the bottom slopes downward on both sides in
the form of a valley. As shown by the contours in Fig.1b, the valley
to the south, which opens into the Ionian Sea, is much more extensive and has a gentler bathymetric slope.

Generally speaking, the Mediterranean has an interface separating Levantine water masses from the Atlantic water masses. In the
vicinity of Messina this interface is situated at a depth of ~ 150m
(Vercelli, 1925; Vercelli and Picotti, 1925). The T-S diagrams in
Fig.6b show the differences in the surface and intermediate layer
water types found typically on either side of the straits. The
resident Tyrrhenian Levantine water is supplied primarily through
the larger Sicilian channels and is only locally modified by contributions through Messina. The T-S values of the Levantine water
found over the Messina sill are intermediate, at $T \simeq 14.2\,°C$,
$S \simeq 38.6\%$, $\sigma_t \simeq 28.94$; those of the Atlantic water are estimated at
$T \simeq 16.6\,°C$, $S \simeq 38.0\%$, $\sigma_t \simeq 27.93$, although, being seasonal, the latter
are more difficult to define. Over the sill, the time averaged
interfacial depth is ~ 30 m, i.e. ~ 120 m above the adjoining
basins. The Tyrrhenian Atlantic water is lighter than the Ionian
(e.g. $\sigma_t \simeq 27.52$, compared with 27.87 at 50 m)   and the mean flow
is into the Ionian at ~ 0.1 m/s. The mean flow in the opposite
direction is the Ionian Levantine water flowing northward into the
Tyrrhenian at ~ 0.13 m/s. A time-averaged velocity profile over the
sill is shown in Fig.7. These speeds can increase up to 0.5 m/s
as a result of local wind forcing. The average hydrological flux
is about $1.3 \times 10^4\,m^3/s$ in both directions. The general hydrological

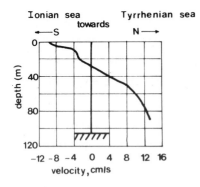

Fig. 7. The vertical profile for the time-averaged currents measured by Vercelli (1925) at his St. 1 on the sill (see also Defant, 1940).

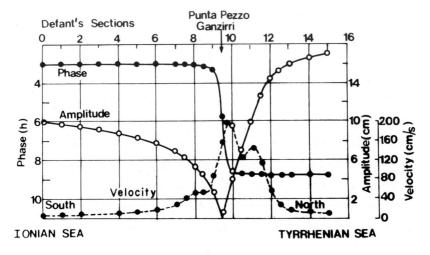

Fig. 8 . Distribution of tidal height, amplitude and phase, and tidal current speed as computed by Defant (1940) through the Strait of Messina.

situation has been discussed by Colacino, Garzoli, Salusti (1980).

The tides of the two main basins of the Mediterranean Sea have rather small amplitudes - 0.10 m south of Messina, 0.14 m north of Messina - but their phase changes radically by 5 hr along the ~10Km through the strait itself (Fig.8). The resulting tidal velocities can attain 2-3 m/s (Vercelli, 1925). Using spectral analysis Vercelli found that 80-90 % of the velocity's energy was due to the tide, the main component of which was the M_2 tide.

These violent currents give rise to intrusions, internal waves, bores, vortices, etc.. Abbate, Dalu, Salusti (1982) have shown that the Reynolds number can be as high as $R \simeq 10^8$ in the ~ 10 Km long region around the sill. Considerable tidal mixing of water masses of different origines (Levantine, upper layers of Atlantic, deeper Atlantic waters) can be observed. Vercelli (1925) estimated a maximum production of mixed water of $\sim 0.3.10^6 m^3/s$.

5-TIME-EVOLUTION OF THE INTERFACE

Defant (1940,1961) based his study of the tides in the Strait of Messina on the results of Vercelli's cruise (1925). In his model the currents are treated as barotropic. In fact, Vercelli's stations over the sill show little baroclinicity, since they are generally homogeneous in water mass and virtually depth-independent in speed. However, on more thorough analysis, the sill is seen to be alternatively occupied by different water masses because the interface separating them displays very large fluctuations (more than 100 m; Griffa, Marullo, Santoleri, Viola, 1982). Hence, Messina is somewhat similar to the better-studied Strait of Gibraltar (Lacombe and Richez, 1982), the difference being that for the Strait of Messina the interface intersects the air-sea surface. With such amplitudes in the interfacial motion it is not surprising that internal waves propagate away from the region.

In accordance with Hopkins, Salusti, Settimi (1982) we shall now investigate the time evolution of the interface on the basis of the Vercelli data, using a simple dynamical balance between the interfacial slope and the acceleration of the baroclinic component. Let us consider a two-layered representation using the equations of motion:

$$u'_t + u'u'_x = -g \, \eta_x \tag{1}$$

$$u''_t + u''u''_x = -g \left(\frac{\rho'' - \rho'}{\rho''} \right) H_x - \frac{\rho'}{\rho''} g \, \eta_x \tag{2}$$

where u', ρ' are the tidal (axial) speed and density in the upper layer, and u'', ρ'' in the lower layer, η is the free surface elevation, H is the interfacial depth and g is the gravitational

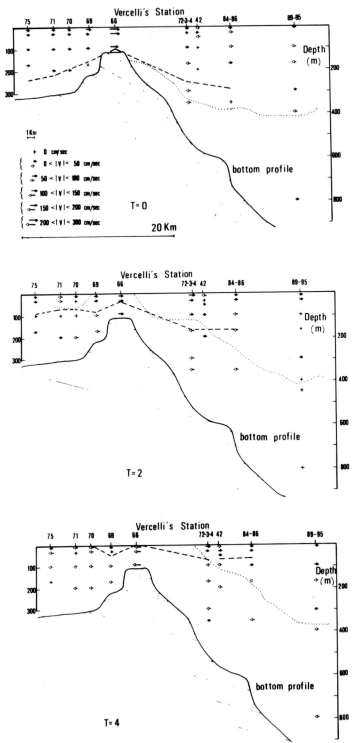

Fig. 9.   The time evolution of the water mass interface over a semi-diurnal cycle. The dashed line is the result of Hopkins, Salusti, Settimi (1983) computations (equation 5), and the dotted line is

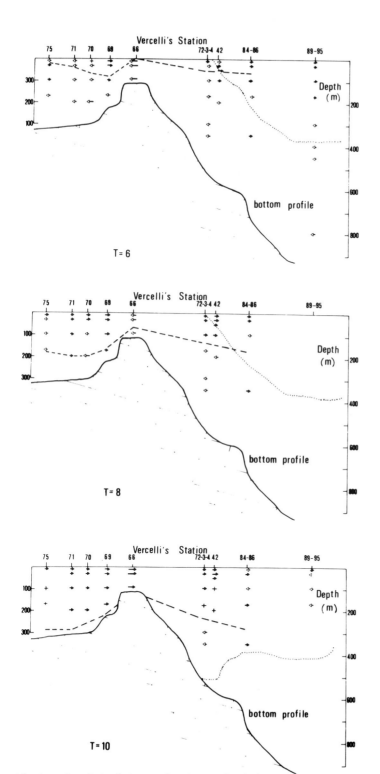

that calculated by Del Ricco (1981). The arrows show the velocities
observed by Vercelli (1925) and the + signs indicate 'no data'.

acceleration. The lateral dimension, along with the Coriolis term, was neglected after a careful analysis had been made of the tidal ellipses (Vercelli, 1925). Although the Coriolis force actually does sometimes lead to the interface having a transverse slope (Colacino et al., 1979), the effect is somewhat mitigated by the selection of central Vercelli stations or cross-channel station averages. To avoid wind or bottom stress contamination, very near surface or bottom records have not been used.

By subtracting equation (2) from (1) we obtain

$$u_t + u'u'_x - u''u''_x = g'H_x \qquad\qquad u \equiv u' - u'' \qquad\qquad (3)$$

where H' is the tidal time-dependent position of H, $g' = \dfrac{\rho'' - \rho'}{\rho''} \cdot g$ and $g'/g'' \simeq 1$. Since the main tidal component is the semidiurnal tide (M_2), the quadratic terms without any simple M_2 periodicity can be disregarded. We can then arrive at the simple relation (Hopkins et al., 1982)

$$u_t = g'H'_x \qquad\qquad (4)$$

With the values of u' and u'' taken from Vercelli's stations, the resulting H' distribution is shown as a dashed line in Fig.9, for approximately every two hours of tide.

Hopkins, Salusti, Settimi (1983) results are comparable with the model calculations of Del Ricco (1981). He used a numerical, viscous, nonlinear, two-dimensional, vertical model based on the formulation used by Hamilton (1971) and Elliott (1976) to study baroclinic currents in estuaries. His boundary conditions were strictly tidal and had only M_2 periodicity. Stratification was represented using the Vercelli (1925) March data set. Del Ricco's (1981) computed interface is shown as a dotted line in Fig.9. The arrows in Fig.9 indicate the directional amplitude of the M_2 tidal velocity as determined by Vercelli. It is apparent that the three different methods agree as far as the general picture of interface time evolution is concerned, although there is some degree of discrepancy.

The edge of the cold spot seen in the thermal images was compared with the distance of the outcropping interface between the two water layers, as computed by Del Ricco (1981, Table 2).

It must also be mentioned that part of the energy of these motions is radiated away as trains of nonlinear internal waves (Alpers, Salusti, 1983; see Fig.10).

TABLE 2
Comparison between observed (by satellite) and theoretical (according to Del Ricco, 1982) distances of the outcropping interface, south of the sill. The residual effect of tidal mixing has been subtracted (Böhm et al. 1983).

TIDAL PHASE (hours)	COMPUTED DISTANCE (Km)	DISTANCE (Km) OBTAINED BY SATELLITE IMAGERY
3	0$\pm$0.5	missing
8	9.4$\pm$0.5	8.$\pm$2
6	10.1$\pm$0.5	10.$\pm$2
5	9.2$\pm$0.5	7.$\pm$2
6	10.1$\pm$0.5	8.$\pm$2
7	9.4$\pm$0.5	5.$\pm$2

DISCUSSION

The physical origin of a large, cold-water patch observable in the Strait of Messina and a long strip along the eastern Sicilian shelf is discussed. Field measurements have shown that the cold strip is due to a peculiar marine water - the C water - flowing southward from the region of the sill of the Strait of Messina. The hydrological characteristics of the C water are $T \simeq 16$-$14\,^{\circ}C$,

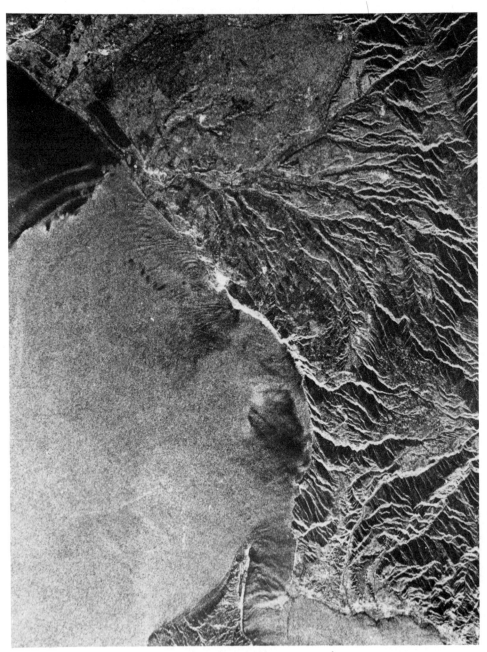

Fig. 10. Digitally processed SAR image from SEASAT orbit 1149 (Sept. 15, 1978; 0817 GMT) showing Calabria, the northeastern tip of Sicily and the Strait of Messina. Packets of circular-shaped internal waves are propagating along the Calabrian coast.

$S \simeq 38.7-38.2 \permil$, $\sigma_t = 28.5-29.0$. Its southward motion follows the sharp Sicilian shelf with velocities of 0.2-0.5 m/s.

This motion is not always regular since eddies and meanders are observable from both satellite imagery and hydrological measurements. During our PRIME '82 cruise, moreover, the C water flow was interrupted by an intrusion of Ionian water.

The above-mentioned strip is generated in the region near the sill of the Strait of Messina. Here, according to F.Vercelli (1925), large tidal currents give rise to a complex, spectacular evolution of the various layers of marine water that flow through the Strait (Levantine water, surface Atlantic water, deep Atlantic water). Vertical excursions of tidal origin (up to 100 meters) have been observed (Griffa et al., 1982), together with trains of nonlinear internal waves. The Reynolds number R can be as high as $R \sim 10^8$. The effect on the air-sea surface of these violent phenomena is to produce a cold patch of the same kind as that observed by satellite. The results obtained by thermal imagery analysis are in fairly good qualitative agreement with our historical and theoretical knowledge.

REFERENCES

Abbate M., Dalu G., Salusti E., 1982. Energy Containing Eddies in the Strait of Messina. Il Nuovo Cimento 5C:571-585.
Alpers W. and Salusti E., 1983. Scylla and Charybdis Observed from Space. Journal of Geophysical Research, Vol.C3:1800-1808.
Böhm E., Magazzù G., Wald L., Zoccolotti L., 1983.(Work in progress)
Colacino M., Garzoli S. and Salusti E.,1980. Currents and Counter-currents in Eastern Mediterranean Straits. Il Nuovo Cimento 4C:123-144.
Defant A., 1961. Physical Oceanography, Vol.II. Pergamon Press 598 pages.
Defant A., 1940. Scilla e Cariddi e le correnti di marea nello Stretto di Messina. Geof. Pura e Appl. Vol.2: 93-112.
Del Ricco R., 1982. A Numerical Model of the Vertical Circulation of Tidal Strait and its Application to the Messina Strait. Il Nuovo Cimento Vol.5C:21-45.
Griffa A., Marullo S., Santoleri R. and Viola A., 1982. Preliminary Observations of Large Amplitude Tidal Internal Waves Near the Strait of Messina. Submitted to Cont. Sh. Dyn.

Hamilton P., 1971. A Numerical Model of the Vertical Circulation
    of Tidal Estuaries and its Application to the Rotterdam Waterway.
    Geophys.J.Royal Astron.Soc. 40: 1-21.
Hopkins T.S., Salusti E. and Settimi D., 1982. Tidal Currents and
    Internal Waves in the Strait of Messina. In press on Journ.
    Geoph. Research.
Lacombe H. and Richez C., 1982. The Regime in the Strait of Gibral-
    tar. Elsevier Oceanographic Series 34:13-73.
Philippe M. and Harang L., 1982. Surface Temperature Fronts in the
    Mediterranean Sea from Infrared Satellite Imagery. Elsevier
    Oceanographic Series 34: 91-128.
Vercelli F., 1925. Crociere per lo studio dei fenomeni nello Stret-
    to di Messina (R.N. Marsigli, 1922-23), Vol.1. Il regime delle
    correnti e delle maree nello Stretto di Messina. Commissione
    Internazionale del Mediterraneo. Venice, Italy.
Vercelli F. and Picotti M., 1925. Crociere per lo studio dei feno-
    meni nello Stretto di Messina (R.N. Marsigli, 1922-23), Vol.2.
    Il regime chimico fisico delle acque nello Stretto di Messina.
    Commissione Internazionale del Mediterraneo, Venice, Italy.

KINETIC STUDY OF SELF-PROPELLED MARINE VORTICES BASED ON REMOTELY SENSED DATA

T. NISHIMURA[1], Y. HATAKEYAMA[2], S. TANAKA[3] and T. MARUYASU[1]

[1]Science University of Tokyo, Noda City (JAPAN)

[2]Asia Air Survey, Co. Ltd., Nurumizu Atsugi City (JAPAN)

[3]Remote Sensing Technology Center of Japan, Roppongi Tokyo (JAPAN)

ABSTRACT

Based on remotely sensed data, a kinetic study is developed about the self-propelled marine vortices. The motion is detected on remote sensing imageries and its mechanism is revealed using the kinematics of the isolated vortices in the perfect fluid. It is shown that the self-propelled marine vortices have an important role to activate the shelf-sea environment around Japan by inducing substantial mass transport and coastal currents.

INTRODUCTION

In the shelf sea around Japan, there are two typical kinds of high speed marine currents. One is a warm oceanic current running along the Pacific Coast, called "the Kuroshio" and another is the tidal current through straits in the Seto Inland Sea.

A recent development in remote sensing has made it practicable to detect isolated marine vortices which are produced by the shear between these high speed currents and the stagnant water of the coastal zone. Also, the motion of these vortices can be often surveyed by remote sensing.

In the spring of 1976, our research group had a chance to start the field survey on the very fast tidal current and the exciting tidal vortices at the Naruto Strait. The tidal current at this strait often reaches 10 knots and causes many exciting tidal vortices. Each vortex usually exceeds 20m in diameter and 2m in water surface depression. Some scientists have tried to survey these rapid flows from boats, but they could not get sufficient data to explain the mechanism of the vortices. In this circumstance, we initiated field surveys of the rapid tidal current by using remote sensing from Landsat and aircrafts (Maruyasu et al., 1981).

Since then, we have always given our careful attention to the shelf sea area on remote sensing imageries from Landsat, aircrafts and NOAA. From these observations, we have found many isolated marine vortices of various scales but we have also surveyed their motion in time series.

The most noticeable motion that has interested us is that some marine vortices move along coast self-propelled with considerable speed. We call these vortices "self-propelled marine vortices", and the kinetic effect of the coast the "image effect". In this report, based on remotely sensed data, kinetic features of these self-propelled marine vortices and their role are discussed by introducing the idea of "image effect".

ISOLATED VORTICES IN THE MARINE FIELD
Isolated vortices along the Kuroshio

In the marine field around Japan, four main oceanic currents are running along its coast. Among them, a warm oceanic current running along the Pacific Coast "the Kuroshio" is the most dominant current. This current supplies a large amount of warm and fresh water from the south, produces a mild climate, a lot of rainfall and rich fishing resources to our country.

As the Kuroshio has a speed of a few knots, it posseses quite a strong inertia and causes a strong shear between itself and the stagnant coast water. When the Kuroshio takes its course close to the coast, its main stream approaches the coast at a range of several tens kilometers offshore (Fig. 1). As the sea bottom in this range is sharply inclined, we can expect at that time a narrow region of strong shear lying between the high speed main current and the coast. The strongest shear is expected especially at the starred portions, where the coastline juts out sharply. In fluid mechanics, this narrow region is called "turbulent boundary layer" and has been one of the main subjects of researh.

In the study of fluid mechanics, flow visualization techniques have often been used in order to estimate the most dominant elements of a flow field, intuitively. It has especially played an important role in the study of the turbulent boundary layer. The development of the wing theory which contributed much to the aircraft technology or that of the coherent structure theory which will be used later in this study are the remarkable results of these techniques.

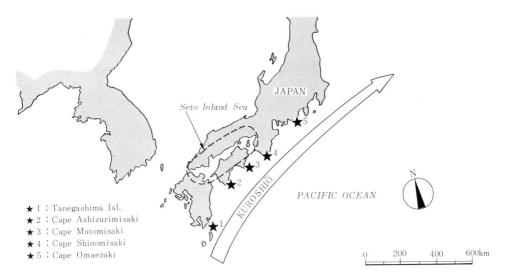

Fig. 1. Outline of navigation course of the Kuroshio. Starred portions are the points of strong shear.

These suggest to us that, if an appropriate flow visualization technique is applicable to our study of the turbulent boundary layer in the macro-scale marine field, we will obtain much visual information about the most dominant element, intuitively.

A recent development in remote sensing has made it possible to offer a flow visualization technique applicable to such macro-scale flow in the natural field. This epoch-making technique can offer essential information not only by visualizing the macro-scale marine flow field in one area, but also by carrying out a quantitative field survey synchronously over a wide area of the sea.

Based on some NOAA imageries, Tanaka et al.(1982) suggested that the shear region between the Kuroshio and the Pacific Coast is filled with isolated vortices which have a deterministic system and a relatively long life. Fig. 2 shows a NOAA imagery analyzed by them, on which we detect some counterclockwise vortices produced in the sea area, lying between the coast and the Kuroshio, especially at the starred portions in Fig. 1. The diameter and the height of these vortices are about several tens kilometers and 1km to 3km.

From the macroscopic view on this NOAA imagery, we observe that these counterclockwise vortices fill up the velocity gap between the high speed current of the Kuroshio and the stagnant water along the coast. They are also expected to contribute well to the horizontal mixing between the Kuroshio water and the coast water.

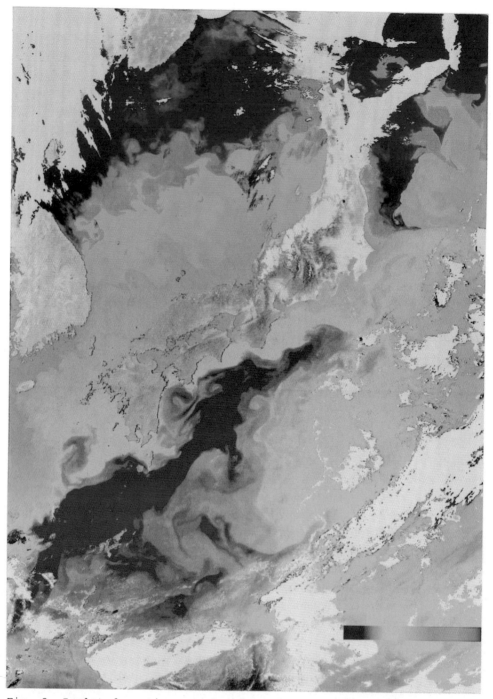

Fig. 2. Isolated vortices produced by the Kuroshio.
(Source: Tanaka, 1982)

These mean that the kinetic features of these isolated vortices are the most dominant element in the turbulent boundary layer between the Kuroshio and the Pacific Coast.

Isolated vortices caused by tidal currents

Another high speed marine current is the tidal current running through the straits in the Seto Inland Sea. Fig. 3 shows the locations of the straits of high speed tidal currents. The highest speed under the spring tide condition is also indicated. It reaches about 10 knots at the Naruto Strait or at the Kurushima Strait. This flow field is also macroscopically visualized by applying remotely sensed data.

Fig. 4 is an airphoto of the southward tidal current running through the Naruto Strait, which was taken at 1,000 meter flight height. The tidal current was running downwards in the center of the water course of 800m in width, with a speed of 9.1 knot. Two free turbulent boundary layers are generated downstreams from the top of the shoals jutting from both sides, and a series of isolated vortices are observed along each boundary layer. These vortices, called here "small scale tidal vortices", have a diameter of about 20m and a height of 30m to 50m.

Fig. 5 is a Landsat MSS 4 imagery of the eastern part of the Seto Inland Sea. On the imagery, we can apparently observe some

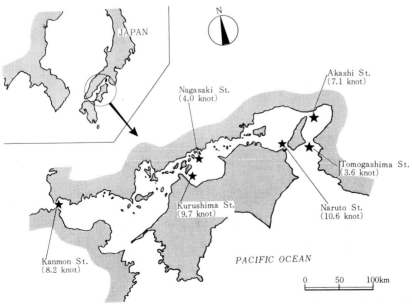

Fig. 3. Straits of high speed tidal current in the Seto Inland Sea.

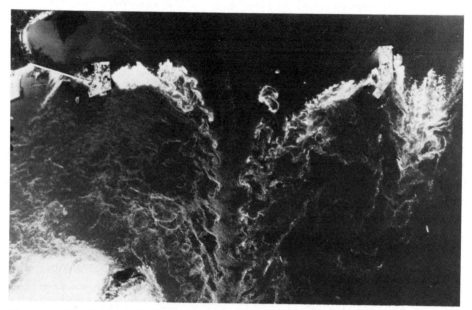

Fig. 4. Airphoto of the tidal current at the Naruto Strait.   Flight height is 1,000m and the current speed is 9.1 knot.

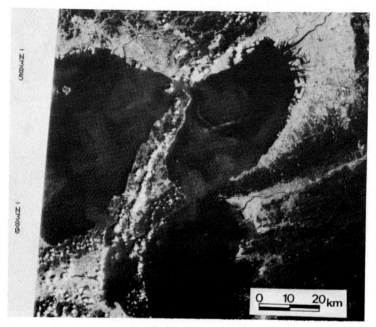

Fig. 5. Landsat MSS-4 imagery of the tidal current in the eastern part of the Seto Inland Sea.

vortices around the straits of Naruto, Akashi and Tomogashima. Their size is about 1km to 2km in diameter and about 50m in height. This type of tidal vortices, called "large scale tidal vortices", are produced after the amalgamation of those small scale tidal vortices over 6 hours, half of a tidal period.

A field study of these tidal vortices at the Naruto Strait was produced by Maruyasu et al. (1981) by applying remote sensing. Airplane remote sensing was applied to small scale tidal vortices. Landsat remote sensing was applied to large scale tidal vortices. The kinetic features of these vortices and their role in the sea environment around the strait were discussed based on remotely sensed data. Especially, the self-propelled motion of the large scale tidal vortices were shown to cause the strong tidal-exchange through the strait between the two open water basins.

From the results, it is concluded that the kinetic features of the isolated vortices is the most dominant element in the fluid mechanics at the strait.

Features of the isolated marine vortices

All of these isolated marine vortices detected on the remote sensing imageries have two kinds of features in common. One is that their space scale is determined by the local topography of the coastline, and another is that each vortex has a fairly small aspect ratio ; the ratio of its height to the diameter.

These features can be transformed as follows by introducing the coherent structure theory. Fig. 6 shows a schematic view of the marine flow field concerned here. The horizontal flow field (a) is divided into three regions : potential flow , turbulent boundary layer and stagnant water region lying along the coast. When a high speed potential flow is running near (along) the coast, a turbulent boundary layer develops in the narrow water region lying between this main current and the coast. The structure of this boundary layer is affected strongly by the topography of the coast. If the coast has a smooth topography, a wall boundary layer grows, that is filled with random turbulent eddies of rather short life. When these eddies are exfoliated from a jutted coast like a cape, they form a free boundary layer. In this free boundary layer, the exfoliated eddies easily form a systematic vortex structure. In this vortex structure isolated vortices have a rather longer life, and grow in steps to larger scale vortices through amalgamation among themselves. In the coherent structure theory, this systematic

(**a**) Plan view

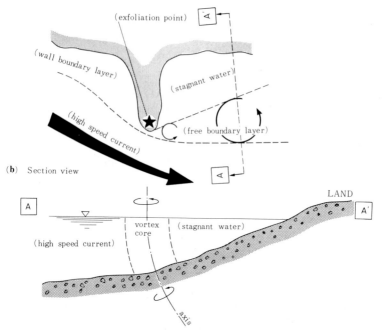

(**b**) Section view

Fig. 6. Schema of isolated marine vortices in the boundary layer.

vortex structure is called "coherent structure".

The coherent structure theory seems to be effectively applied to the fluid mechanics of the boundary layer in the sea. This is because the theory has an ability to explain the mechanics of the turbulent boundary layer based on the kinematics of isolated vortices composing it (Davies & Yule, 1974). In our study where the object is to examine, at a macroscale, the marine flow field, remote sensing is expected to play an essential role as a flow visualization technique that is necessary to apply that theory.

According to the coherent structure theory, each isolated marine vortex is a line vortex, and the Helmholtz's theorem reveals that a line vortex must terminate on the boundary of the flow field or must connect itself to form a closed loop. Because of the small aspect ratio in our case, each isolated marine vortex must be a line vortex whose axis connects the sea surface directly to the sea bottom as in Fig. 6. Also because of the very small aspect ratio, its kinetic features are under the strong influence of the bottom features. The "image effect" discussed in this study is an example of such kinetic influences of the sea bottom.

REMOTE SENSING

What is "image effect" ?

   The idea of the "image‚effect" is based on the well known kinematics of isolated vortices in perfect fluid. The following discusion is developed from Lamb's textbook (1932) which is a typical textbook of fluid mechanics of perfect fluid.

   Now, let's assume a vortex-pair composed of two vortex cores having equal and opposite circulation as in Fig. 7(a). Then, the vortex-pair moves from left to right with uniform speed:

$$U_p = \frac{\Gamma}{4\pi a} \tag{1}$$

where, U_p : moving speed of the vortex-pair.
      Γ : vortex strength defined as the circulation.
      a : distance between the two vortex axes.

   Fig. 7(a) shows the steady streamline pattern, when viewed from the coordinate system shifting together with the vortex-pair. The corresponding stream function ψ is steady and is represented by:

$$\psi = -\frac{\Gamma}{2\pi}\left(\frac{y}{2a} + \log\frac{r_1}{r_2}\right) \tag{2}$$

In this equation, r_1 and r_2 indicates the distance between the concerned point and the vortex axes. In Fig. 7(a), the shaded portion of the closed streamlines indicates the "carrier" of the substantial fluid mass that is transported together with the vortex-pair. The oval geometric configuration of the outer brim of the carrier is described as a streamline:

$$\psi = 0 \tag{3}$$

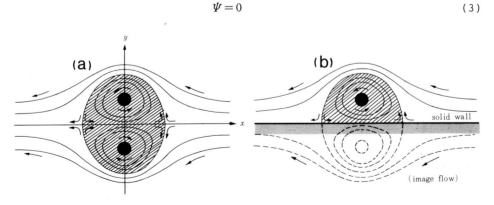

Fig. 7. Schema of the "image effect". The core is pasted and the carrier of substantial mass transport is shaded.

Under the assumption of perfect fluid flow, we can place a solid boundary along any planes which are not crossed by any of the fluid movement. Therefore, a stationary solid boundary wall along the symmetry plane as in Fig. 7(b) can be placed without creating any effects. Then, the vortex core must propel itself parallel to this solid boundary wall as if an "image vortex-pair" actually exists on the back sides of the solid wall. The speed of this self-propelled vortex core is calculated by eq.(1) where, a is the distance between the vortex axis and the boundary wall. This self-propelled vortex is followed by a substantial mass transport whose carrier is the shaded portion in Fig. 7. Here, it must be noticed that this carrier transports not only the vortex core itself but also the surrounding fluid mass that has no vorticity. In this study, this kinetic effect of a solid wall to a neighboring vortex is called the "image effect".

An interesting image effect study in the aerial field relating to the aircraft turbulence was conducted by Barker & Crow (1977). When an aircraft takes off, a pair of trailing vortices are left over the ground as is illustrated in Fig. 8(a). Barker et al. analyzed the mechanics of these trailing vortices by simplifying them to a vortex-pair as shown in Fig.8(b). That is, at the start the vortex-pair propels itself downward to the ground under its own dominant propelling effect, and in a little while, the pair separates and each vortex moves outwards receiving the image effect from the ground surface. Barker et al. called this image effect the "ground effect".

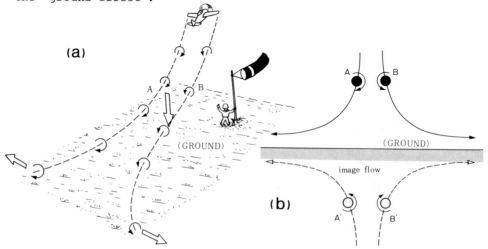

Fig. 8. Schema of the "ground effect". (a) trailing vortices. (b) Barker's analysis.

Self-propelled vortex produced by the Kuroshio

A self-propelled vortex produced by the Kuroshio was detected on Landsat and NOAA imageries by Hatakeyama et al.(1981) at the mouth of the Suruga Bay.

The site is the sea area around Cape Omaezaki which was shown as a starred portion in the chart of Fig. 1, and is under the strong friction effect of the Kuroshio. Fig. 9 shows the depth chart, in which we see that the Shoal Kanesunose juts out abruptly into the Pacific Ocean and extends over 30km offshore from Cape Omaezaki. Another shoal extends into the Pacific Ocean from Cape Irozaki, the top of the Izu Peninsula. Sandwiched by these two shoals, Suruga Bay which is a deep and wide bay opens to the Pacific Ocean through a water course of about 40km in width.

Some field surveys in the past pointed out that the bay water is strongly influenced by the intrusion of the Kuroshio water when the Kuroshio takes its course close to the coast. This means that a water exchange is accelerated at the baymouth under this condition, however, its mechanism has not been investigated well until now.

On a Landsat imagery obtained at 9:41 on Oct. 22nd 1979, a flow pattern with a distinct isolated vortex was observed at the mouth of the Suruga Bay. Fig. 10 shows the Band-4 image of that scene. On this imagery, we find an apparent flow pattern across the bay mouth, that may be formed by the discharge from the Tenryu River.

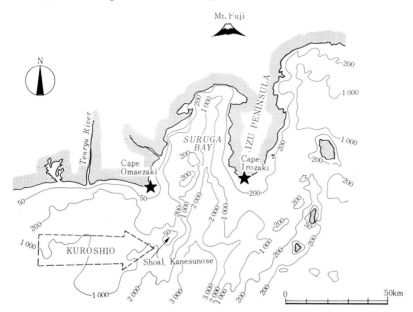

Fig. 9. Bottom features of the site.

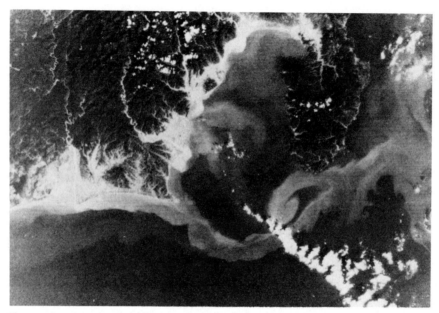

Fig. 10.   Landsat MSS-4 imagery at 9:41, 22-OCT-1979.

The remarkably lucky occurence of two natural phenomena at almost the same time allowed us to obtain the clear flow pattern on the Landsat imagery. One reason for this luck was that the Kuroshio was taking its course close to the coast when this Landsat imagery was obtained. Another was that since the Typhoon-20 hit Japan at that time, a lot of rainfall was supplied to this site about 3 days before. Consequently, a lot of turbid water discharged from the Tenryu River during the afternoon of Oct. 19th served as a natural tracer of the flow visualization.

Based on these backgrounds, the Landsat imagery was interpreted from the viewpoint of fluid mechanics. Here, the Kuroshio and the Shoal Kanesunose correspond to the high speed current and to the jutted topography, in the schematic flow field of Fig. 6. The results of the interprtetation are as follows.

"The turbid water discharged from the Tenryu River flows eastwards along the Pacific Coast which has a very smooth topography. This flow forms a wall boundary layer along the coast whose width is about 20km and is smaller than the length of the Shoal Kanesunose. Therefore, the wall boundary layer exfoliates at that shoal and

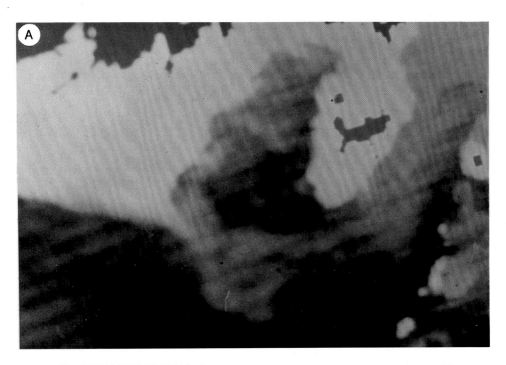

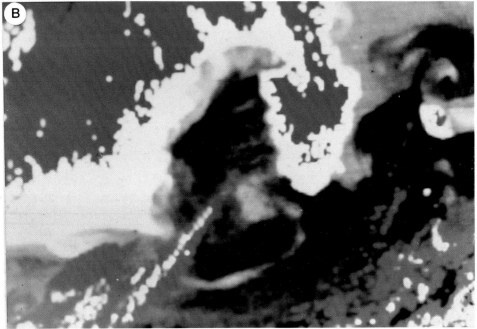

Fig. 11. NOAA imageries. (Source: Hatakeyama, 1981)
(A) at 18:26, 20-OCT-1979. (B) at 07:44, 21-OCT-1979.

forms a free boundary layer. This free boundary layer runs across the baymouth to reach Cape Irozaki, filling up the velocity gap between the Kuroshio and the stagnant bay water. The counterclockwise vortex near Cape Irozaki is formed by the amalgamation of the circulation flux exfoliated from the Shoal Kanesunose."

Based on the coherent structure theory, it is apparent that the kinetic features of that counterclockwise vortex holds a key to reveal the fluid mechanics at the mouth of the Suruga Bay. But, its motion cannot be interpreted directly on the Landsat imagery, because it gives only a temporal state of the vortex motion.

In order to survey the kinetic features of this isolated vortex as a time series, four sets of NOAA data were analyzed effectively. These data were those obtained at 18:26 on Oct. 20th, at 19:44 on Oct. 21st, at 08:05 on Oct. 22nd and at 07:44 on Oct. 23rd. Photos of Fig. 11 are the first and the last of these NOAA imageries and Fig. 12 shows each state of the isolated vortex interpreted on the respective NOAA imageries. Based on these imageries and the maps, the following results were obtained.

It is verified that the isolated vortex was produced exactly at the Shoal Kanesunose. Its formation process spreading over 25 hours is shown on the imageries (A) and (B). The vortex strength Γ is estimated under the assumption of circulation flux conservation by:

$$\Gamma = \Gamma_{flux} \cdot \Delta t = \frac{1}{2} U_{main}^2 \cdot \Delta t \fallingdotseq 3 \times 10^4 \mathrm{m^2/sec} \qquad (4)$$

where, Δt : period required for the vortex formation, about 25 hours in this case.

Γ_{flux} : circulation flux from the Shoal Kanesunose.

U_{main} : current speed of the main stream about 3km/hour in this case.

It is noticeable that this vortex formation process is followed by an excellent entrainment of the surrounding water into the vortex as shown in Fig. 12(a). This process is supposed to promote the water mixing at the baymouth and to refresh well the bay water.

The self-propelled motion can be apparently observed on Fig. 12. The vortex, having completed its formation process (State-(B)), moves across the baymouth toward Cape Irozaki (State-(C)), is self-propelled along the west coast of the Izu Peninsula and is trapped into the bay (State-(D)). This self-propelled motion is supposed to be caused by the image effect of the coast and of the shoal stretching out from the Izu Peninsula.

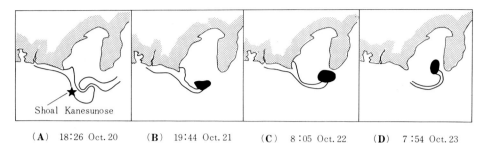

(**A**) 18:26 Oct. 20   (**B**) 19:44 Oct. 21   (**C**) 8:05 Oct. 22   (**D**) 7:54 Oct. 23

Fig. 12. Time series of the motion of the vortex.
(Source: Hatakeyama, 1981)

Self-propelled vortex produced by tidal currents

Fujiwara (1979) found that at the Akashi Strait a large scale
tidal vortex has a much noticeable kinetic feature. He found this
through airphotos or Landsat imageries and described the motion of
the tidal vortex. In this section, we show that this motion is
a self-propelled motion induced by an image effect.

The site is the sea area around the Akashi Strait whose location
was shown as a starred portion of Fig. 3. That is, it is under the
strong friction effect of the tidal current. As this strait is a
water course that connects the Seto Inland Sea to the Pacific
Ocean, the tidal-exchange through it contributes well to refresh
the water environment of the inland Sea. The most contracted water
course is about 3.7km in width and is about 100m in water depth.
Through this strait, the tidal current runs up and down with a
period of about half a day. The highest speed reaches about 6 knot
in the extreme case of a full moon.

Such a high speed tidal current produces systematic large scale
tidal vortices in the sea area around the strait. On the Landsat
imagery of Fig. 5, we can observe a large scale tear-drop water
mass in the central part of the Osaka Bay, which is revealed later
to be composed of large scale tidal vortex produced at the
Akashi Strait.

We can observe a self-propelled vortex on the Landsat imagery of
Fig. 13, which was obtained at 10:06 on Oct. 24th 1972. The tidal
flow condition at that date is based on the Tide Table as follows:

            05:23 --- slack from east to west
            08:42 --- westward flow maximum 6.4knot
            12:09 --- slack from west to east
That is, this Landsat imagery indicates the tidal flow field at the

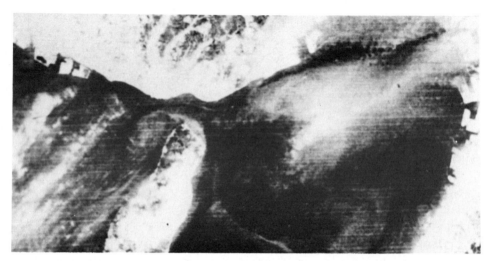

Fig. 13. Landsat MSS-4 imagery at 10:06, 24-OCT-1972.

Akashi Strait at the state 2 hours before the slack from west to east. On the imagery, we can observe that the wall boundary layer along the east coast of Awaji Island is exfoliated at the north top of the island and is forming a large scale counterclockwise vortex. Because of its self-propelled motion towards the strait, this vortex is not driven away by the westward main current and its diameter reaches the same scale as the width of the strait.

Fujiwara (1979) described the kinetic features of large scale tidal vortices by a schema of Fig. 14 which shows the flow fields at the three typical stages of the tidal current conditition. He found this through the analysis of Landsat imageries and airphotos. Each stage corresponds to the condition of (a): westward current, (b): slack from west to east and (c): eastward current. Fujiwara's description is as follows.

Stage-(a) : Westward current. A counterclockwise vortex is produced at the top of the Awaji Island, that is clearly identified as a turbid water mass boiled up. In the Osaka Bay, the current into the strait is composed of two branch streams. The large scale tear-drop water mass in the central part of the Osaka Bay is the water mass which has been transported through the strait by the previous eastward current. This water mass moves far into the Osaka Bay counter to the westward current surrounding it.

Stage-(b) : Slack from west to east. The counterclockwise vortex formed at the previous stage is pushed into the strait at

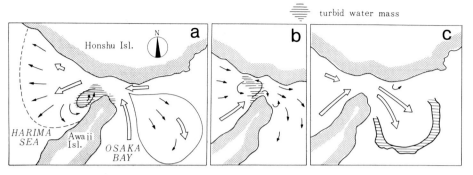

Fig. 14. Schema of tidal flow field at the Akashi Strait (Source: Fujiwara, 1979). (a) westward flow. (b) slack. (c) eastward flow.

the head of the eastward current.

Stage-(c) : Eastward current. The water mass having passed through the strait forms a large scale tear-drop water mass. The counterclockwise vortex is broken and stretched into a narrow band of turbid water which extends along the outer brim of the tear drop water mass.

In this description, the most noticeable phenomenon is the motion of the counterclockwise tidal vortex, and we can surely assume that it shows the self-propelled motion induced by the image effect of the coast of Awaji Island.

An air survey of the tidal flow field at the Akashi Strait was performed by the Honshu Shikoku Bridge Authority in 1976. Fig. 15 shows the examples of the airphotos taken on Oct. 26th 1976, on which two typical stages of the formation process of the counter-clockwise vortex are visualized with a lot of tracers of about 1m square on the sea surface. Airphoto-(A) was taken at 08:59 and airphoto-(B) was taken about two hours later at 10:42. Fig. 16 shows the schema of the flow field which was deduced from these airphotos. In the figure, the condition of the tidal current at the respective shutter chance is also shown.

The flow field is divided into three portions ; the main stream, wall boundary layer along the coast of the Honshu Island and that along the coast of the Awaji Island. The latter boundary layer exfoliates at the top of the island, and forms a free boundary layer. The circulation in this free boundary layer amalgamates to form a counterclockwise vortex. Stage-(A) is the beginning of this formation and Stage-(B) shows the midst of it. During these, the vortex grows larger and its diameter finally reaches the scale of the width of the strtait.

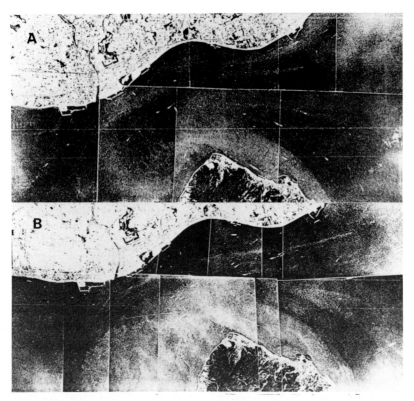

Fig. 15. Sequent airphotos on 26-OCT-1976 (Source: Honshu Shikoku Bridge Authority). (A) at 08:59-09:06. (B) at 10:42-10:49.

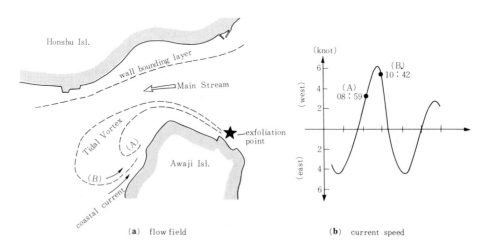

(a) flow field

(b) current speed

Fig. 16. Schema of the tidal flow field. (a) flow field. (b) time variation of the current speed.

Applying the idea of the image effect, the motion of the tidal vortices in Fig. 14 is represented as follows. At the Stage-(a), a counterclockwise vortex is formed. As this vortex propels itself northwards under the image effect of the coast of the Awaji Island, it is not driven westwards into the Harima Sea under the action of the westward main current. It grows larger staying at the same place, and when the westward current finishes (Stage-(b)), is pushed into the most contracted part leading to the eastward water mass flux. This motion is also caused by the image effect. At Stage-(c), the counterclockwise vortex is destroyed and stretched along the outer brim of the tear-drop water mass which is formed by the water mass transported eastwards following the counterclockwie vortex. At the outlet to the Osaka Bay, both coastlines have a point of strong inflection and the boundary layer exfoliation advances easily. Therefore, a large scale tidal vortex-pair is formed in the Osaka Bay, which is the natural shape of the tear-drop water mass. As the vortex-pair propels itself into the Osaka Bay by the mechanism stated before, the water mass composed of the counterclockwise vortex stays in the Osaka Bay together with the tear drop water mass during the next stage of the westward current.

IMAGE EFFECT THEORY

Vortex-pair model

A sharply inclined bottom is often observed in the coastal sea areas which directly face the high speed marine current. Under this condition, the "vortex-pair model" is constructed in Fig. 17.

Figure 17a shows a vortex-pair. Under the assumption of perfect fluid flow, three solid boundary planes α , β , γ can be put stationary in the flow field without any effects because all of these planes are not crossed by any fluid particle movements.

Figure 17b shows the schema of the "image effect" based on the vortex-pair model. In this model, we assume that the bottom is a horizontal plane and the coast is a vertical cliff. Under this geometric condition, it is reasonably assumed that the isolated marine vortex forms a straight line vortex. Then, the flow field of this marine vortex is equivalent to that of Fig. 17(a) sliced off by the three solid boundary planes. Here, the planes α , β and γ correspond respectively to the sea surface, vertical coast and the horizontal sea bottom. As the results, the marine vortex propels itself along the coast, as if the image flow field existed actually on the back sides of the boundary planes.

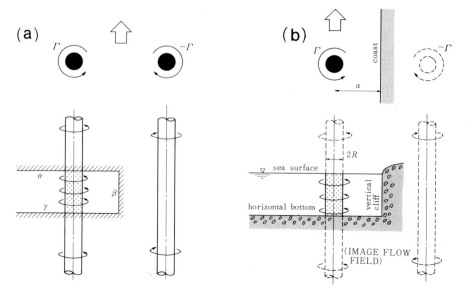

Fig. 17. Vortex-pair model. (a) vortex-pair. (b) image effect.

Fig. 18 shows the streamline pattern of the flow field at the sea surface when we apply this vortex-pair model to the image effect in the sea, which is viewed from the coordinates system moving together with the vortex core. The kinetic features of the vortex core and the surrounding water mass is explained as follows.

  (i)   When a marine vortex is located near the coast, it propels itself along the coast under the image effect.

 (ii)   When we observe this self-propelled marine vortex from the coast, its motion is rightwards for a counterclockwise vortex and is leftwards for a clockwise vortex.

(iii)   The self-propelled speed of the vortex core is estimated by eq. (1) where, a is the distance between the vortex axis and the coastline.

 (iv)   This self-propelled motion is followed by a substantial water mass transport.

  (v)   The carrier of this mass transport is represented as the shaded portion of Fig. 18. It takes the form of a half of an oval, and extends over the distance $2a$ offshore and over the distance $4a$ along the coastline.

 (vi)   The transport of the core can be explained by Helmholtz's theorem but that of the surrounding water mass is explained only by introducing the image effect.

(vii)   Considerable coastal current is induced along the coastline

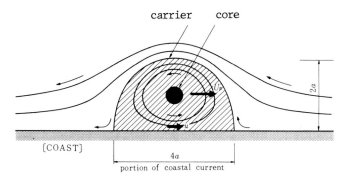

Fig. 18. Streamline pattern based on the "vortex-pair model".

over the range of about $4a$, and its maximum speed is calculated by:

$$U_c = 2 \times \frac{\Gamma}{2\pi a} = \frac{\Gamma}{\pi a}$$
(5)

These are the essential features of the kinematics of a self-propelled marine vortex and the water mass surrounding it.

Vortex-ring model

A more advanced "vortex-ring" model is proposed here in Fig. 19, in which the sea bottom is assumed uniformly inclined. This topographic condition seems more realistic. Fig. 19(a) shows the vortex-ring having a ring radius a, core radius R and the vortex strength Γ. Then, this vortex-ring propels itself with the uniform speed:

$$U_r = \frac{\Gamma}{4\pi a} \log \frac{6a}{R}$$
(6)

along its ring axis. Under the assumption of perfect fluid flow, we can put two stationary solid boundary walls α and β in the flow field without any effect. In this case, these planes contain the ring axis on its surface and are not crossed by any fluid particle movement.

Fig. 19(b) shows the schema of the "image effect" based on the vortex-ring model. In this model, we assume that the sea bottom is uniformly inclined from the shoreline. Under this condition, it is reasonably supposed that the marine vortex forms a part of an image vortex-ring having its ring axis exactly on the shoreline. The radius of this image vortex-ring a is equal to the distance

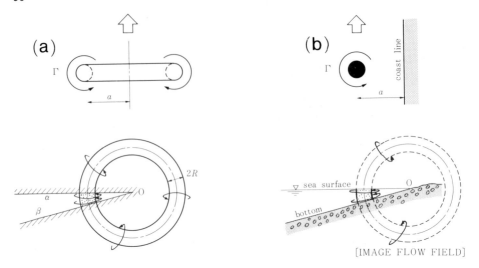

Fig. 19. Vortex-ring model. (a) vortex-ring. (b) image effect.

between the vortex axis and the coast. Then, the flow field of this marine vortex is equivalent to that of the vortex-ring in Fig. 19(a) sliced out by the two solid planes. Here, the planes α and β correspond respectively to the sea surface and to the uniformly inclined sea bottom.

As a result, the marine vortex must move self-propelled along the coast as if the image flow field of the vortex-ring actually existed. The uniform self-propelled speed is calculated by eq. (6), where, a is the distance between the vortex axis and the coast. It must be noticed here that the self-propelled speed in this case is also a function of the core radius R, which was not reflected in the vortex-pair model. When the core radius is very small, the self-propelled speed becomes infinite.

When we view it from the coordinate system shifting together with the vortex core, the streamline pattern takes the form shown in Fig. 20 [drawn with reference to Prandtl and Tietjens (1957)]. In the vortex-ring model, the streamline pattern is strongly influenced by the finite core radius, because the self-propelled speed is a function of it. When the parameter R/a is large (Case-(a)), the carrier reaches the coast and we can expect an induced coastal current. When this parameter is small (Case-(c)), the carrier is confined near the vortex core and we expect no coastal current. The critical parameter, corresponding to the tear drop carrier (Case-(b)), is about 1/100.

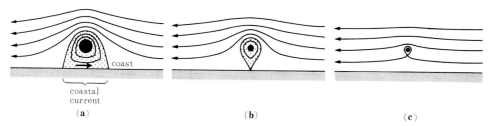

Fig. 20. Streamline pattern based on the "vortex-ring model".

MODEL EXPERIMENT TO VERIFY THE THEORY
Apparatus and methods

Vortex production. To verify the theory, a model experiment was performed. Fig. 21 shows the experimental apparatus. A simplified model of the sea bottom, a flat bottom with a uniform slope 1/20 was constructed in the water basin of 20m x 30m in space. To produce a scaled model of the isolated marine vortex, a rectangular board of 1.8m width is placed vertically on the slope and then is moved towards the coastline, keeping its original position. The board is pulled up from the water surface, then an isolated vortex is left behind in the test area as shown in Fig. 22. The strength and the size of the isolated vortex are controlled by the speed and the stroke of the displacement of this board. The aspect ratio of the vortices is about 1/10, and the ratio R/a is about 1/3. These parameters are of the same order as those often observed for the self-propelled vortices in the sea, and the geometric similarity is satisfied in this model experiment.

Flow visualization. To measure the kinetic parameters, the flow field of the vortex core and its surroundings is visualized by a small floating tracer. On Fig. 22, we see the tracer surrounding the core forming the carrier of the substantial mass transport. The self-propelled speed, distance from the coast and the speed of the coastal current are measured with a measuring tape and a stopwatch.

Underwater measurement. To measure the rotating velocity, a propeller type velocity meter is set in the course of the vortex axis as shown in Fig. 21. As the propeller is set at 5cm water depth pointing perpendicular to the coast, the velocity meter starts initially with negative values and changes to positive values when the vortex axis passes the measuring point. The velocity

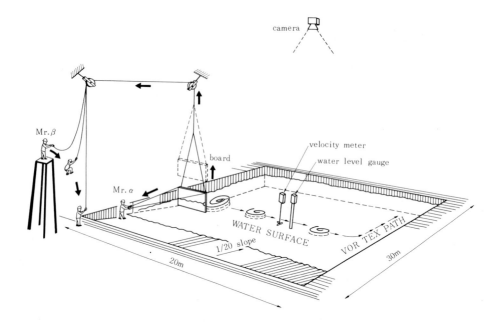

Fig. 21. Apparatus and methods.

Fig. 22. Snapshot of the experiment.

vector is the resultant of the rotating component perpendicular to the coast and the self-propelled speed parallel to it. As the propeller can pick up the first component only, the velocity meter represents the spacial distribution of the rotating velocity around the vortex axis. From these data, we can estimate the vortex strength or the core radius.

To measure the water surface depression along the centerline of the vortex, a capacity type water level gauge is also set stationary in the course of the vortex axis as shown in Fig. 21. These data are also useful to calculate the vortex strength or the core radius. The data from these two sensors are recorded by a pen recorder as a time series.

Results and analysis

Self-propelled speed. The core radius and the vortex strength were calculated from sequent photos, velocity distribution and water surface depression. These data were used to calculate the self-propelled speed by the vortex-ring model.

In each case, two kinds of vector maps of the velocity were obtained from the displacement of the floc of tracers, measured on the sequent photos taken at 5 sec intervals. That is, a vector map viewed from the stationary coordinate system and one from the moving coordinate system. Fig. 23 shows examples of the vector maps of the latter type. It is noticed that these are similar to the velocity field of Fig. 20a and 20b, which are obtained from the vortex-ring model. The vortex core interpreted on the photos

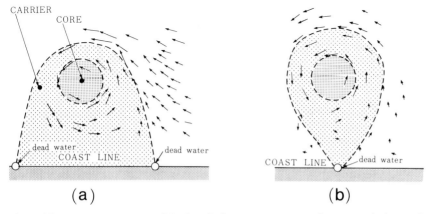

Fig. 23. Vector maps obtained from sequent photos. (a) oval carrier and coastal current. (b) tear drop carrier and no coastal current.

94

as the core of solid rotation is constructed on these vector map. From these vector maps, the first kind of experimental parameters were measured, the core radius R_1 and the rotating velocity u_1 on the circumference of the core.

Fig. 24 shows an example of the record of the outputs from the velocity meter and the water level gauge. The curve (a) shows the rotating velocity distribution. The point of zero velicity is the time when the vortex axis passed the measuring point. The outer brim of the vortex core is indicated on this curve as the two points α and β of the maximum and minimum rotating speed. The linear distribution of the velocity between these confirms the rigid rotation of the core. From this velocity distribution, the second kind of experimental parameters were measured as follows. The measured values R_2 of the core radius was calculated by:

$$R_2 = \frac{1}{2}(U_0 \cdot \Delta t)$$ (7)

where, U_0 : observed value of the self-propelled speed.

Δt : time lag between the points α and β , needed by the vortex axis to pass over the measuring points.

The measured value u_2 of the rotating speed was calculated by:

$$u_2 = \frac{1}{2}(u_\alpha - u_\beta)$$ (8)

The curve (b) shows the water surface depression along the centerline of the vortex. The point of the maximum depression indicates the time when the vortex axis passed over the measuring

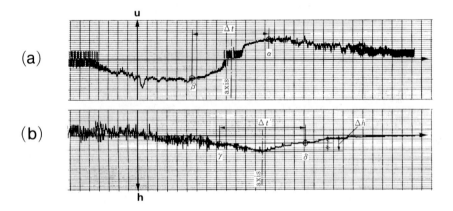

Fig. 24. Outputs from under water sensors. (a) rotating velocity distribution. (b) water surface depression.

point. Under the Rankine Vortex model, the outer brims of the vortex core are indicated as the two points γ and δ of the depression $\Delta h/2$. Also, from this surface depression, the core radius and rotating speed were measured as follows. The measured value R_3 of the core radius was calculate by:

$$R_3 = \frac{1}{2}(U_0 \cdot \Delta t')$$
(9)

where, U_0 : observed value of the self-propelled speed.

$\Delta t'$: time lag between the points γ and δ , needed by the vortex axis to pass over the measuring points.

The measurted value u_3 of the rotating speed was calculated under the Rankine Vortex assumption by:

$$u_3 = \sqrt{g \cdot \Delta h}$$
(10)

From these three kinds of measured parameters, the corresponding calculated value U_i of the self-propelled speed was estimated by:

$$U_i = \frac{\Gamma_i}{4\pi a} \log \frac{6a}{R_i}$$
(11)

based on the vortex-ring model. Here, the parameter a is the observed value of the distance from the coast, and the vortex strength Γ_i is estimated from the respective measured values by:

$$\Gamma_i = u_i \times (2\pi R_i)$$
(12)

Fig. 25 shows the comparison between the observed data of the self-propelled speed and the calculated value from the three kind of data. The ordinate shows the observed value in the respective experimental case and the abscissa shows the calculated value. When the points are plotted precisely on the broken lines, they support the applicability of the vortex-ring model. We find that the calculated value from the photos agrees well with the observed data. However, the value calculated from the outputs from the two underwater sensors gives smaller magnitudes than the observed data. This trend results from the experimental difficulty in setting the sensors precisely on the path of the vortex axis. When these facts are put together, we conclude that the estimation of the self-propelled speed by the vortex-ring model is reasonably applicable to the isolated vortices.

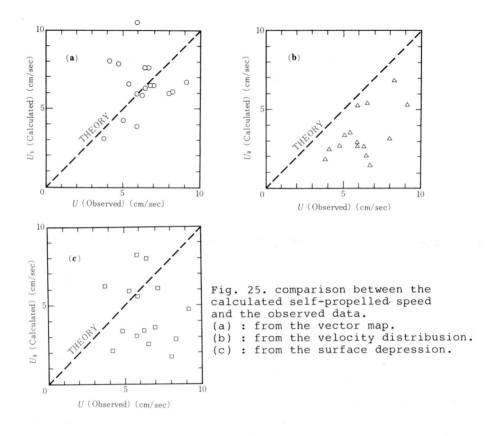

Fig. 25. comparison between the calculated self-propelled. speed and the observed data.
(a) : from the vector map.
(b) : from the velocity distribusion.
(c) : from the surface depression.

Substantial mass transport and coastal current. It was theoretically predicted that the substantial mass transport caused by the self-propelled vortex is composed of two elements : transport of the vortex core and that induced by the image effect itself. The first element is easily observed on the succesive photos in which the crowded tracer staying inside the vortex core moves from left to right, rotating and keeping the original relative position.

Another element that is induced by the image effect itself is detected on the vector maps of Fig. 23. In this experiment, two typical types of vector maps were observed. In the vector maps, the painted portion shows the vortex core observed on the photos and the shaded portion shows the carrier. Its outer brim is drawn as a streamline extending from the dead water points located on the coastline. The vector map (a) has two dead water points on the coastline, and the carrier has a shape of an oval. In this case, a considerable coastal current is observed. The vector map (b) has only one dead water point and has a shape of a tear-drop. In this case, no coastal current is observed.

MODEL EXPERIMENT TO ESTIMATE THE EFFECT OF EARTH ROTATION

Apparatus and methods

Vortex production. To estimate the effect of the earth rotation on the isolated marine vortices, another series of model experiment was performed in a small water basin that was set on a turntable as shown in Fig.26. The turntable is a round flat table with diameter of 2m and is rotated counterclockwise at 2/3 r.p.m., corresponding to the earth rotation on the northern hemisphere. A water basin of 1.5m square was set on the table. The experiment was performed on a horizontal sea bottom and on a uniformly inclined sea bottom with a slope 1/20. The method of vortex production is similar to the former experiment, but in this case. the vortex maker is controlled more closely by a motor.

Similarity. The geometric similarity was estimated by the aspect ratio of the isolated vortex and by the parameter R/a . The vortex produced has a diameter of about 10cm and a height of about 1cm. Its distance from the coast is about 20cm. Then, its aspect ratio is about 1/10, and the parameter R/a is about 1/4. These are the order often observed for the prototype marine vortex.

The kinetic similarity was estimated by a dimensionless time defined by:

$$T^* = \frac{T_v}{T_e}$$

(13)

where, T_v : period of the vortex rotation.

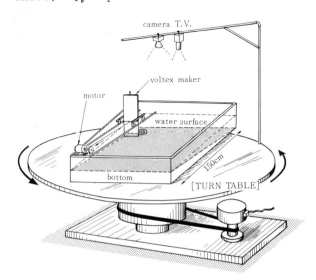

Fig. 26. Apparatus and methods.

T_e : period of the earth rotation.

The higher this parameter is, the stronger the effect of the earth rotation becomes. In this experiment, the rotation period of the vortex is about 10sec and that of the earth is 90sec. Therefore, the dimensionless time for this model is:

$$(T^*)_m = \left(\frac{T_v}{T_e}\right)_m \doteqdot \frac{10\,\text{sec}}{90\,\text{sec}} \doteqdot \bar{O}\,(0.1) \tag{14}$$

For the prototype marine vortices observed, this parameter is estimated as follows. That is, for the vortex produced by the Kuroshio at the mouth of the Suruga Bay:

$$(T^*)_p = \left(\frac{T_v}{T_e}\right)_p \doteqdot \frac{10\,\text{hour}}{\left(\dfrac{24\,\text{hour}}{\sin 40°}\right)} \doteqdot \bar{O}\,(0.1) \tag{15}$$

and for the tidal vortex produced at the Akashi Strait:

$$(T^*)_p \doteqdot \frac{0.3\,\text{hour}}{\left(\dfrac{24\,\text{hour}}{\sin 35°}\right)} \doteqdot \bar{O}\,(0.01) \tag{16}$$

Therefore, the vortex produced in this experiment corresponds more suitably to the isolated marine vortex at the Suruga Bay produced by the Kuroshio.

Flow visualization and measurements. During the production, the vortex core was visualized by a blue dye. The locus of the vortex core was recorded by a camera in a time series. The rotation speed of the vortex core and its deformation process were also recorded. The vortex strength was calculated from the rotating speed and the core radius, which were measured on the photos.

Results and analysis

Case of horizontal sea bottom. Following three cases were examined:

Case-(A) : counterclockwise vortex & no earth rotation

Case-(B) : counterclockwise vortex & earth rotation

Case-(C) : clockwise vortex & earth rotation

Fig. 27 shows the kinetic features of the isolated vortex in each experimental case, on which the respective sequent photos were taken at 24sec intervals. The locus of the vortex core movement traced on these photos are shown in Fig. 28.

From these data, the followings results are obtained.

Case-(A) : This case was tested for the basis of the estimation. After its production, the vortex propels itself parallel to the

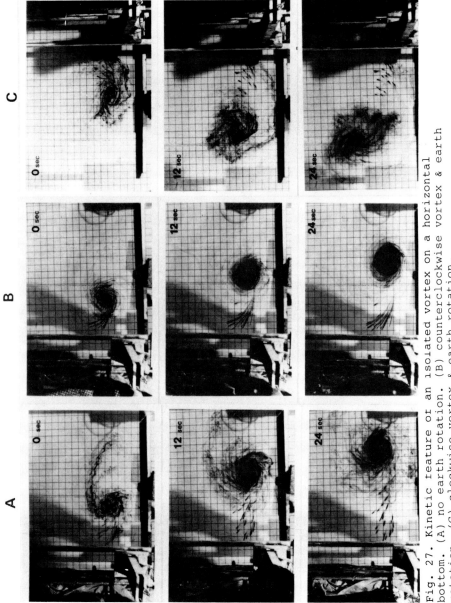

Fig. 27. Kinetic feature of an isolated vortex on a horizontal bottom. (A) no earth rotation. (B) counterclockwise vortex & earth rotation. (C) clockwise vortex & earth rotation.

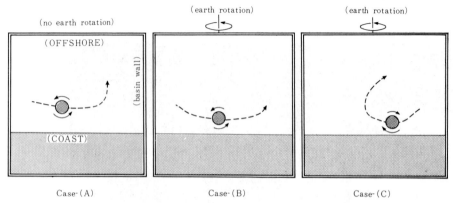

Fig. 28. Locus of an isolated vortex on a horizontal sea bottom.

coast under the image effect of the coast. The observed self-propelled speed and the calculated value by the vortex-pair model show good agreement.

Case-(B) : In this case also, the vortex moves parallel to the coast and the self-propelled speed is almost the same as Case-(A). Here, it is noticeable that the vortex can keep itself more stable for a long time in the self-propelled motion. It can maintain its original form more easily than in the case of no earth rotation.

Case-(C) : In this case, the vortex moves along the coast only at the start, and then, it propels itself offshore. It must be noticed here that the clockwise vortex deforms very rapidly.

From the results, we conclude that the image effect is accelerated by the earth rotation in the case of the counterclockwise vortex on the northern hemisphere but is made powerless in the case of the clockwise vortex.

Case of uniformly inclined sea bottom. In this experiment, those three cases were also examined and the following results were obtained.

Case-(A) : After the production, the vortex propels itself parallel to the coast. The self-propelled speed calculated by the vortex-ring model and that measured directly on the sequent photos agree well.

Case-(B) : In this case, the vortex propels itself towards the right under the image effect of the sea bottom and at the same time it approaches the coastline. The self-propelled speed is approximately the same as observed in Case-(A) but the motion towards the coast is more accelerated by the effect of the earth rotation

together with the effect of the inclined sea bottom. The strong
stabilizing effect is observed.

Case-(C) : In this case, self-propelled motion parallel to the
coastline is not found. The vortex stays in the sea area where it
was produced and is deformed very rapidly.

From these results, we conclude that the kinetic influence of
the earth rotation in this experiment is approximately equivalent
to that observed on the horizontal sea bottom. The effect of the
bottom inclination is revealed in the acceleration of the vortex
movement **towards** the coast or   offshore. This effect should
be caused by the action of the relative vorticity that is induced
following the change of the water depth.

APPLICATION OF THE THEORY
Vortices by the Kuroshio

The self-propelled speed of the vortex at the Suruga Bay can be
estimated with the vortex-ring model. The vortex strength Γ is
estimated to be about 30,000 m^2/sec by eq. (4). The core radius
and the distance from the coast are measured directly on the NOAA
imageries to be about 7km and 15km. From these parameters, the
self-propelled speed is calculated to be 1.5 km/hour by eq.(6).
This value agrees well with 1 km/hour measured directly on the
sequent NOAA imageries of (C) and (D) in Fig. 12.

The water temperature chart of Fig. 29 shows the temperature
distribution on Oct. 23th of 1979, the day the NOAA imagery (D)
was obtained. This chart was published by Dr. Nakamura who is
studying the short time fluctuation of the oceanic condition of

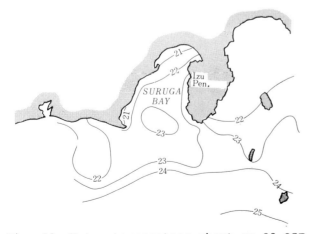

Fig. 29. Water temperature chart on 23-OCT-1979 (Source: Nakamura).

this sea area (Nakamura et al. (1981)). In this chart, we notice
an isolated water mass of closed contour of 23 °C trapped in the
central part of the Suruga Bay. This temperature corresponds to
that of the narrow water region stretching along the baymouth,
therefore, this isolated water mass is considered to be transported
into the bay following the self-propelled motion of the isolated
vortex that was detected on the NOAA imageries. A long and narrow
water zone with slightly higher temperature along the west coast of
the Izu Peninsula is supposed to be transported by the northward
coastal current induced by this isolated vortex.

If these phenomena can be explained by the image effect theory,
a large portion of sea water in the Suruga Bay will be replaced by
the water of the Pacific Ocean in a few days after the formation of
an counterclockwise vortex at the Shoal Kanesunose. This mechanism
is expected to contribute well to the short time fluctuation of the
oceanic condition of the Suruga Bay. For example, Nakamura et al.
indicates that the short time fluctuation has a great influence on
the fishery in this area.

Such a mechanism is also expected to have a dominant role in
refreshing the sea water in other areas located along the Pacific
Coast of Japan (see Fig. 30). In this figure, the corresponding

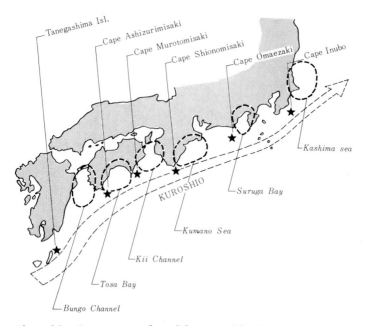

Fig. 30. Sea area of self-propelled marine vortices and those
exfoliation points.

exfoliation points of the circulation are also indicated. The counterclockwise vortices exfoliated at these points have the kinetic feature of moving against the direction of the Kuroshio, propelled by themselves under the image effect. As a result, they are trapped into these sea areas and refresh the sea water environment by promoting the substantial mass transport and coastal currents. Based on our experimental results, it is estimated that these self-propelled vortices are stabilized and have a longer life-time as the result of the earth rotation.

Tidal vortices at the Akashi Strait

The self-propelled speed of the counterclockwise vortex at the Akashi Strait can be estimated by the vortex-ring model as follows. On the airphotos of Fig. 15, the core radius and the distance from the coast are estimated to be about 1km and 2km. The current speed along the outer brim of the vortex core is considered to be equal to the speed of the main current of 2 m/sec at that time. Then, the vortex strength is calculated by:

$$\Gamma = 2 \times (2\pi R) = 2 \times 2 \times \pi \times 10^3 \fallingdotseq 1.2 \times 10^4 \mathrm{m}^2/\mathrm{sec} \qquad (17)$$

The self-propelled speed calculated by eq. (6) is about 10 m/sec and this is more than the speed of the main current, therefore, this counterclockwise vortex is not pushed away by the inertial force of the westward flow.

At the slack from west to east, this vortex moves self-propelled into the most contracted part of the strait under the image effect. Following this motion, a substantial mass transport occurs which is visualized as the turbid water shown in Fig. 14. This substantial mass transport contains the surrounding water mass around the vortex core, a portion of which comes from the Harima Sea along the west coast of the Awaji Island. That is, we conclude that this vortex plays an important role in the tidal-exchange between the Harima Sea and the Osaka Bay. Fujiwara (1979) also commented that this mass transport is sometimes followed by a transport of fry of sand launce from its spawning ground in the Harima Sea.

The counterclockwise vortex is expected to induce a considerable coastal current along the west coast of the Awaji Island. This coastal current has an efficient influence on the tidal flow field at the strait under the slack condition from westward to east. Fig. 31(b) schematically shows the time variation of the spacial distri-

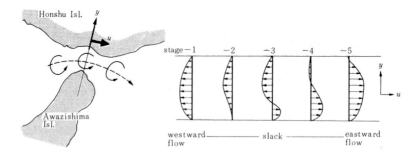

Fig. 31. Influence of the counterclockwise tidal vortex on the reflux process at Akashi Strait.
(a) transit of the vortex. (b) reflux process.

bution of the eastward velocity component at the most contracted section in the figure (a), which is obtained based on the general description of the flow field in the Tide Table. In the figure, we can indicate that the tidal flow starts to change its direction eastwards initially at the top of the Awaji Island. This reflux region gradually extends and finally all the section is covered by the eastward flow region. This reflux process takes about 30 minutes to complete, and this period is considered to correspond to the period being necessary for the counterclockwise vortex to pass through the Akashi Strait.

CONCLUDING REMARKS

Based on remotely sensed data, a kinetic study was developed about the self-propelled marine vortices under the image effect of the coastline.

On a Landsat imagery, an isolated vortex was found to be produced at the mouth of the Suruga Bay and its self-propelled motion was surveyed quantitatively by sequent NOAA imageries. On Landsat imageries and airphotos, a counterclockwise tidal vortex at the Akashi Strait was found to be propelled by itself along the coast and its formation process was surveyed by sequent airphotos.

To explain the mechanism of this self-propelled motion, a kinetic theory labeled "image effect theory" was proposed. Based on this theory, the self-propelled speed can be calculated by a function of the core radius, the distance from the coast and the vortex strength. It was also theoretically revealed that the self-propelled marine vortex is followed by substantial mass transport and a coastal current of considerable speed. These theoretical results were confirmed by a larger scale model experiment and the effect of the earth rotation was estimated by a smaller scale model experiment using a turntable.

It was revealed that the self-propelled marine vortices have an important role in refreshing the shelf-sea environment. That is, the sea water is activated considerably by these self-propelled marine vortices followed by substantial mass transport and coastal currents.

ACKNOWLEDGEMENT

The image analysis and the interpretation of NOAA data are the result of joint development among the authors and Mr. Toshiro Sugimura of the Remote Sensing Technology Center of Japan.

During the course of this research, the authors were greatly supported by Dr. Koichi Kinose of the National Research Institute of Agricultural Engineering, and by the following students of the laboratory of hydrodynamics; Mr. Yuichi Ogihara, Mr. Kenichi Azuma, Mr. Toyoji Joko, Mr. Shigeo Murai and Mr. Yoichi Kitajima.

The authors wish to express many thanks to these coworkers.

Financial support was provided by the Ministry of Education, Science and Culture under the Grant in Aid of Scientific Research, (No.56550356).

REFERENCES

Barker, S.J. and Crow, S.C., 1977. The motion of two dimensional vortex-pairs in a ground effect. J. Fluid Mech., Vol.82, part 4, pp. 659-671.
Davies, P.O.A.L. and Yule, A.J., 1975. Coherent structure in turbulence. J. Fluid Mech., Vol. 69, part 3, pp. 513-537.
Fujiwara, T., 1979. Water mass from the Harima Sea trapped into the Osaka Bay. Proc. of Autumun Meeting of Oceanogr. Soc. of Japan, pp. 12-13. (in Japanese)
Hatakeyama, Y., Tanaka, S. and Nishimura, T., 1981. A formation process of an oceanic vortex analyzed by multi-temporal remote sensing. Proc. of 15th ERIM, pp. 1173-1185.
Lamb, H., 1932. Hydrodynamics. 16th edition, Cambridge Univ. Press.
Maritime Safety Agency of Japan, annually. TIDE TABLES, VOLUME I, NIPPON AND ITS VICINITIES.
Maruyasu, T., Onishi, S. and Nishimura, T., 1981. Study of tidal vortices at the Naruto Strait through remote sensing. Bull. of the Remote Sensing Laboratory, No. 1, Science Univ. of Tokyo.
Nakamura, Y. and Chiba, H., 1981. Short-term fluctuation of oceanic conditions at the adjacent waters of northern part of Izu Island. Bull. of Shizuoka Pref. Fish. Exp. Stn., No. 15, pp.1-7.
Prandtl, L. and Tietjens, O.G., 1934. Fundamentals of Hydro- and Aeromechanics. (Dover Edition, 1957)
Tanaka, S., Sugimura, T., Nishimura, T. and Hatakeyama, Y., 1982. Accuracy of direct measurement of mean water surface velocity of the Kuroshio using multi temporal NOAA-6 imageries, Proc. of 16th ERIM. (to be published)

STUDY OF VORTEX STRUCTURE IN WATER SURFACE JETS BY MEANS OF REMOTE SENSING

SOTOAKI ONISHI

Civil Engineering Dept., Science University of Tokyo, Noda City, Chiba pref. Japan, 278.

ABSTRACT

The fluid dynamic characteristics of water surface jets in nature are investigated by means of several remote sensing techniques, including aerial photography, infrared photography, colour sonor and the Landsat data. The tidal currents in the Naruto strait in Japan are selected as the object of the study, because these currents present typical features of water surface jets. The dynamic characteristics of the jets to be discussed include coherent structure of vortices in shear layers, structure of potential core, circulation, vertical profiles of the vortices, preferred mode of the jet, and behaviour of the jets in the far field.

INTRODUCTION

The fluid dynamic characteristics of jets are one of the most interesting and one of the most studied topics related to turbulent flow. But almost all existing results have been obtained on the basis of either theoretical or experimental laboratory investigations. In this paper, the deterministic features of surface water jets in nature are investigated by means of several remote sensing techniques such as aerial photography, infrared photography, the Landsat data as well as colour sonor.

The tidal currents in the Naruto strait in Japan will be used as a prototype for the study. Onishi and Nishimura (1980) and Maruyasu, Onishi and Nishimura (1981) have previously investigated the tidal currents in the same strait from the point of view of mass exchange through the strait. They showed that vortices in the currents should play an important role in these phenomena. In this paper, the tidal currents are considered as water surface jets; the fluid dynamic structure of the vortices is shown to behave as an unstabilized perturbation in the shear layers in the flow development zone.

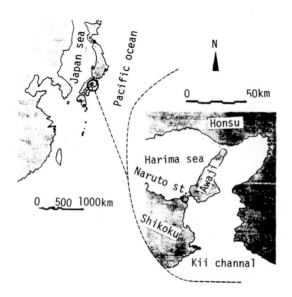

Fig. 1. Location of the Naruto strait

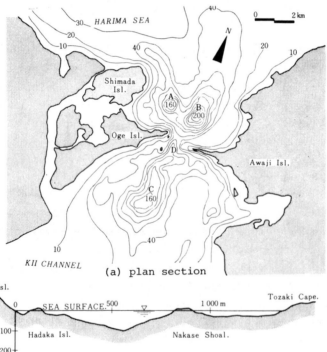

(a) plan section

(b)   section view

Fig. 2. Bottom configuration of the Naruto strait

COHERENT STRUCTURE OF VORTICES IN SHEAR LAYERS AND POTENTIAL CORE
Structure of shear layer in turbulent jet

The Naruto strait is located in the eastern part of the Seto
inland sea (Fig. 1), the coastal zone of which is one of the most
industrialized area in Japan. Fig. 2 shows the sea bottom configuration around the strait. The width of the strait is about 1,000
meters and the maximum depth is more than 80 meters. Tidal level
on each side of the strait ebbs and rises periodically with a
period of about 12 hours as shown in Fig. 3; during spring tides,
the difference in sea level reaches approximately 1.5 meters and
causes velocity of about 10 knots.

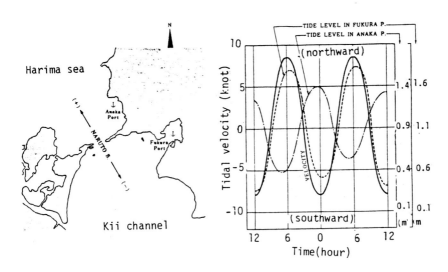

Fig. 3. Time variation of tidal current within Naruto strait and
sea level on both sides of the strait.

Photo 1 is an example of an aerial photograph of the tidal
current in the strait from an altitude of about 1,000 meters,
representing a stage of the southward tidal flow of 9.1 knots flowing from Harima sea into the Kii channel. In turbulent theory,
jets have often been discussed by dividing the flow into a zone of
flow development and a zone of developed flow as indicated in
Fig. 4. In the Naruto strait, a typical example of the flow development zone is shown in Photo 1, in which series of vortices
exist along a pair of shear layers. Although such shear layers are
considered in stochastic turbulent theory as consisting of many
vortices of various diameters, those in Photo.1 spread out over

110

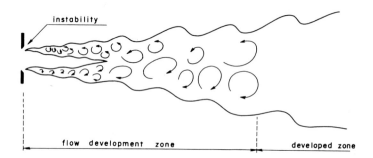

Fig. 4. Development of a turbulent jet.

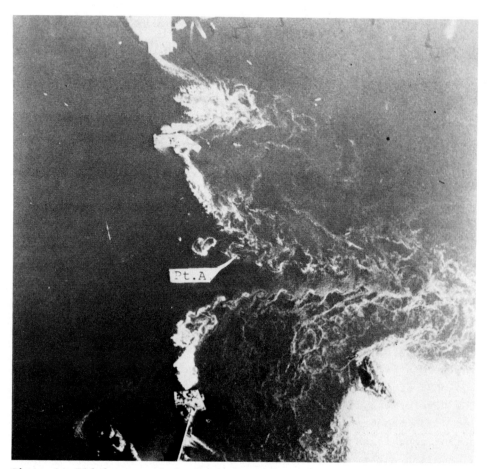

Photo 1. Tidal current in Naruto strait (Photographed at 1000 m height).

Photo 2.   Coalescence of vortices.

the whole width of the shear layers and have a coherent structure.
It has been said that such turbulent vortices displaying an organized structure are formed through the rapid coalescence of
two or more neighbor vortices of smaller scale. Photo 2 is an
example of aerial photograph obtained at a lower altitude (600
meters) which shows more clearly the coalescing vortices.

Structure of the potential core

Another interesting aspect of the flow development zone is the
structure of the potential core. This aspect was studied by infrared photography. Photo 3 (a)-(e) are thermal images taken at an
altitude of 600 meters at the following times on July 16, 1977 :
(a) 11:11 a.m, i.e., 30 minutes before the maximum southward flow
condition; (b) 11:16 a.m; (c) 11:25 a.m; (d) 11:38 a.m; and (e)
11:46 a.m. Darker parts in the image indicate lower temperature
of the water surface. The flow is visualized by the mass of warmer
water passing through the strait and one can see the potential core
being divided into two regions along the centre of the main flow,
denoted as Front A in photo 3 (d). One can also see many small,
dark, circular spots in the potential core. As shown most conspisciously in photo 3 (a), those circular spots, whose diameters are
about 100 meters, are distributed in a row along a line near the
centre of the flow. The author presumes that those spots may be
billows formed through deformation process of perturbed waves
added on the shear layer as indicated schematically in Fig. 5.
Scorer (1978) derived the following condition for instability of
such waves in a stratified fluid :

$$k > k_c = \frac{2g\beta}{\omega^2 \Delta z} \tag{1}$$

where k denotes the wavenumber, g the acceleration of gravity,
ω the vorticity, Δz the thickness of the vorticity layer and β the
density gradient represented by

$$\beta = \frac{\Delta\rho}{\rho\Delta z} . \tag{2}$$

Denoting the velocity difference across the vortex layer by ΔU,
the vorticity is represented by

$$\omega = \frac{\Delta U}{\Delta Z} . \tag{3}$$

Fig. 5. Deformation process of perturbed waves on the shear layer.

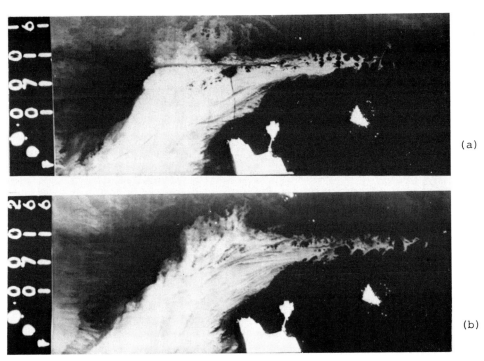

(a)

(b)

Photo 3.

114

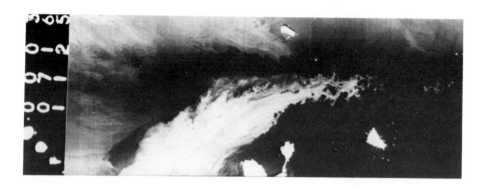

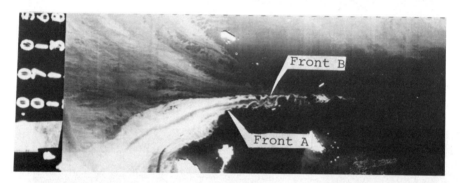

Photo 3. Thermal images of water surface during southward flow (flight height of 600 m).

The substitution of eq. (2) and (3) into eq. (1) yields

$$k > k_c = \frac{2g \ \Delta\rho}{\rho \ (\Delta U)^2} \ . \tag{4}$$

Hence, the critical wavelength L_c for the instability is given by

$$L_c = \frac{2\pi}{k_c} = \frac{\pi \rho (\Delta U)^2}{\Delta \rho \cdot g} . \tag{5}$$

The waves of wavelength smaller than L_c are unstable.
In the case $\Delta \rho = 0$, i.e., in a homogeneous fluid, all perturbed
waves shall be unstable. When $\Delta U = 0$ in a non-homogeneous fluid, k_c
becomes infinitely large, that is, all waves are stable. In the
Naruto strait a reasonable value of $\Delta \rho / \rho$ is 0.005.
In Photo.3(a) we estimate the spatial interval between neighbouring
spots at about 50 meters. Assuming that the interval length is
equal to the critical wavelength, the corresponding velocity
difference ΔU_c is estimated by eq. (5) as follows :

$$\Delta U_c = \left\{ \frac{L_c \cdot \Delta \rho / \rho \cdot g}{\pi} \right\}^{1/2} \underset{\sim}{} 0.88 \text{ m/sec.} \tag{6}$$

Although the velocity difference across the front A was not
measured, one may consider that the value of 0.88 m/sec is quite
likely, taking into account that the mean tidal velocity under
maximum flood condition is about 4 m/sec. Under the condition of
$\Delta U = 0.88$ m/sec, the circular spots in the stage described in
Photo.3(a) can be stable. Indeed, those small circular spots
behave just like rollers in the shear layers of a turbulent flow.
In the stages (b) to (f) of Photo.3, the current velocity increases
and the corresponding critical wavelength also increases.
Therefore, the series of circular spots observed in Photo.3(a)
should become more widely spaced.

FIELD OBSERVATION OF VORTEX STRENGTH
    Referring to Fig. 6 and assuming that the strength of vorticity
generated at point P is preserved during the downstream transpor-
tation along the shear layer, the vorticity flux passing through
a section I in a unit time can be estimated as follows :
with U denoting the velocity in the direction of the x-axis, the
vorticity is

$$\omega = dU/dz,$$

and the vorticity flux through section I, ω_{flux}, is given by

$$\omega_{flux} = \int_{-\infty}^{\infty} \omega \, U \, dz = \int_{-\infty}^{\infty} \frac{dU}{dz} \, U \, dz = \frac{1}{2} \, U^2. \tag{7}$$

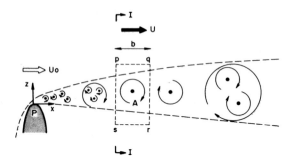

Fig. 6. Schema of the coalescing vortices

Denoting the velocity of propagation of the vortex by c and the interval between neighbouring vortices by b, the time interval, τ, of the vortex becomes

$$\tau = \frac{b}{c} . \tag{8}$$

Therefore, the vorticity transported by the vortex, namely, the vortex strength Γ is represented by

$$\Gamma = (\omega_{flux}) \, \tau = \frac{1}{2} \, (U^2) \, \frac{b}{c} . \tag{9}$$

Then, if U, b, c are measured in the field, the vortex strength can be estimated by the above equation.

Another way to estimate the vortex strength is a method based on the water surface depression at the vortex centre. Assuming a line vortex as shown in Fig. 7, the vortex strength can be represented by

$$\Gamma = 2R \sqrt{g.\Delta h}, \tag{10}$$

where, R denotes the radius of the rotational flow region adjacent to the vortex axis, g, the gravitational acceleration and Δh, the water surface depression at the vortex centre.

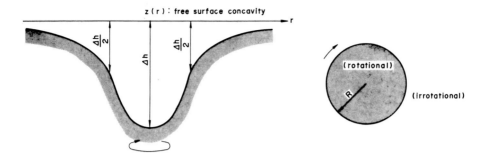

Fig. 7. Rankine's vortex model

With the above relation, one can obtain the vortex strength if the
values of R and Δh are measured in the field.

The spatial intervals between vortices can be observed easily by
ordinary aerial photography. The velocity of propagation of the
vortex and the current velocity can be obtained by a series of
aerial photographs taken at fixed time intervals or by application
of the principle of the Kameron effect. Maruyasu, Onishi and
Nishiura (1981) tried such observations in the Naruto strait with
a time interval of three seconds, on April 1st 1977, July 2nd 1977,
March 8th 1977 and February 24th 1978, respectively.

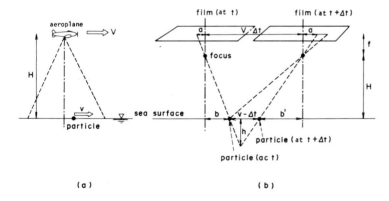

Fig. 8. Principle of the Kameron effect

The current velocity and the velocity of propagation can be also
measured by application of the Kameron effect as follows : in
essence, suppose that a water particle at the surface flows with a
velocity v as shown in Fig. 8 in the same direction as that of the
airplane; then that velocity v can be represented by the equation

118

$$v = V - \frac{1}{\Delta t} \frac{H}{f} (a + a'),$$ (11)

where H is the flight height, V, the flight speed, Δt, the time
interval between photographs, f, the focal distance of the camera;
the definitions of a and a' are indicated in Fig.8. So, when viewed
through a stereoscope, the displacement of the water particle in
the time interval Δt produces the illusion of being a relative
height h. Therefore, the velocity of the water particle v can be
observed through a stereoplotter. Hence, contour maps showing the
velocity distribution can be obtained. Fig. 9 is an example of
such a map.

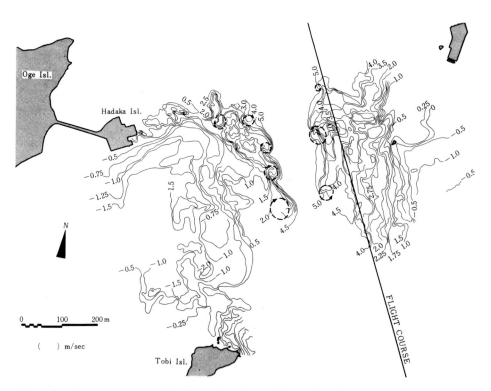

Fig. 9. Contour map of the velocity distribution

Next, the water surface depression Δh in eq.(10) can be measured
by photographing the water surface from a pair of airplanes
synchronously. In this study, the surface was photographed in the
Naruto strait from two airplanes at an altitude of 1,000 meters

Date	Vortex No.	U (m/sec)	X (m)	b (m)	c (m/sec)	τ (sec)	Γ (m²/sec)
1977 Apr. 1st	1	3.7	100	17	2.1	8	60
	2	—	120	28	3.1	9	60
	3	—	160	63	4.2	15	100
	4	—	240	90	3.8	24	160
	5	—	340	96	4.6	36	240
	6	—	440	163	4.0	41	280
	1′	—	170	73	3.5	21	140
	2′	—	240	74	2.9	25	270
	3′	—	320	66	2.1	32	220
	4′	—	380	86	3.3	26	180
	5′	—	490	129	4.6	28	190
	6′	—	630	143	4.2	34	230
1977 Jul. 2nd	1	4.1	150	80	3.9	21	170
	2	—	230	74	4.0	19	160
	3	—	300	66	3.5	19	160
	4	—	360	80	3.3	24	200
	5	—	460	96	4.5	21	180
1977 Mar. 8th	1	4.4	100	58	2.5	23	220
	2	—	160	59	3.4	17	170
	3	—	220	64	2.8	23	220
	4	—	290	73	2.5	29	280
	5	—	360	114	4.1	28	270
	1′	—	160	58	2.9	20	190
	2′	—	220	76	2.8	27	260
	3′	—	310	70	3.0	23	230
	4′	—	360	84	2.7	31	300
	5′	—	480	123	3.0	41	390
1978 Feb. 24th	1	4.55	120	44	2.8	16	160
	2	—	160	52	4.2	12	130
	3	—	220	86	4.7	18	190
	4	—	340	82	3.3	25	260
	5	—	390	66	3.5	19	190
	6	—	470	100	4.0	25	260
	1′	—	280	56	4.0	14	140
	2′	—	340	50	3.2	16	160
	3′	—	380	48	2.8	17	180
	4′	—	430	76	3.4	23	230
	5′	—	530	100	5.3	19	200

TABLE 1.

Strength of the vortices observed in the Naruto strait.
The presence (absence) of a prime on the vortex number indicates
that the vortex was observed in the left - (right-) hand free
boundary layer. The distance between the vortices and their generation point is denoted by X.

with an 80 percent overlap to get a series of pairs of synchroni-
zed aerial photographs. Each pair of aerial photographs was analyzed with the A-7 autograph and a contour map of the sea surface
was produced. Fig. 10 is an example of such a map, in which the
surface depressions at the centre of the vortices can be measured.
From those field observations and using eq.(9) or eq.(10), the
vortex strength can be calculated. Details of these estimations
have been previously reported by Onishi and Nishimura (1980).
Therefore, only the final results are presented here in Table 1.
The strength of the vortices in the Naruto strait is
$(1-3) \times 10^2$ m^2/sec, and the velocity of the tidal current in the
narrowest section of the strait is about 4 m/sec. It should be
pointed out that the values of vortex strengths reported in
Table 1 are based on eq.(9). Equation (10) gives a strength of
about 150 m^2/sec when using a vortex radius of 8 meters together
with a water surface depression of about 1.0 meters, both of which
are obtained from Fig. 10.

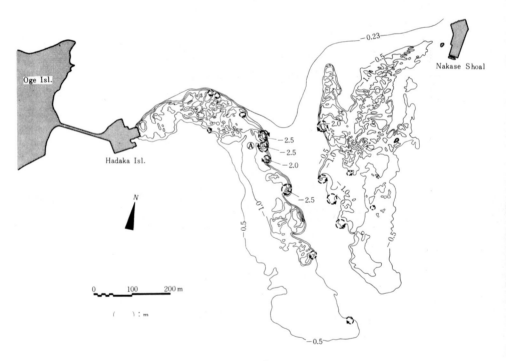

Fig. 10. Contour map of the sea surface

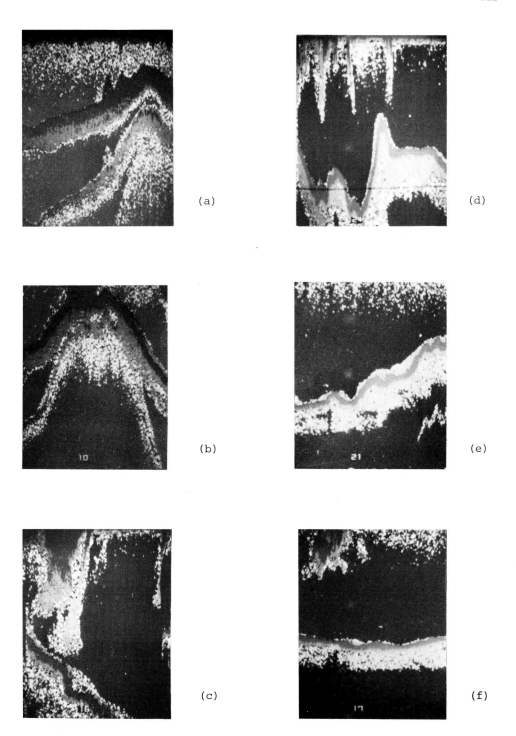

Photo 4. Vertical profiles of the vortices.

VERTICAL PROFILES OF THE VORTICES

Field observation of the vortex profile

Although the vertical profile of the vortex is an interesting
subject to observe in nature, none of the conventional experimental methods has succeeded to do it. The application of the colour
sonor seems to be a way to overcome the difficulty. The principle
is as follows : acoustic waves emitted into the water body are
reflected at the sea bottom and by air bubbles or suspended particles in and around the vortices and the difference in reflectance
shows up as a difference in colour. Photo 4 presents such results
obtained in a southward flow. Panel (a) shows the profile of a
vortex observed in a region of about 14 meters depth with the shear
layer to the right of the downstream direction. Panel (b) shows a
profile observed in another region, about 8 meters deep, downstream
of (a); panel (c) shows observations further downstream. Photo 4
(d) displays the results obtained along a course traversing the
current from Shikoku to Awaji, in which the maximum water depth is
about 22 meters. When the water depth is shallow and the vortex
strength is large enough, the vortices reach the sea bottom, as
shown in Photo 4 (b) and (c). On the other hand, if the water depth
is large enough (as in the case of Photo 4 (d) or if the vortex is
weak (Photo 4 (e)), the vortices do not reach the bottom.
In Photo 4 (a), one can see the vortices stretching downwards
from the water surface. One can also see a double trace of the sea
bottom, the lower one being a result of a secondary reflection of
the acoustic waves. The interesting thing is that there is a
wedge-like configuration visible on the first image of the bottom
topography, but not on the second image. This means that the
reflectivity of the wedge-like bodies is lower than that of the
solid sea bottom, and therefore the author considers these bodies
as indicating upwelling currents associated with the vortices.
These will be discussed later in more detail. To induce such an
upwelling current, it is essential that the vortex reach the sea
bottom. Then, owing to the bottom friction, the tangential velocity is decelerated in the neighbourhood of the bottom; this gives
rise to a secondary flow which is directed radially inwards and
then turns upwards (Schlichting (1968)).

Hydraulic model tests of the upwelling current

Onishi and Nishimura (1980) performed hydraulic model tests to observe the behavior of the upwelling current caused by a vertical line vortex and estimated the upward volume flux associated with the vortices in the Naruto strait.

In a test basin of 0.6 m width and 25 m length, a uniform, free-surface flow of velocity equal to 0.7 cm/sec was generated. The vertical line vortex was produced as follows : a plate, 0.3 m wide, was put suddenly into the uniform flow keeping one edge touching the side wall of the test basin at right angle. It was kept in the uniform flow until a vortex of a certain strength formed at the other edge of the plate. Then the plate was lifted out suddenly and the vertical line vortex remained in the uniform flow. Both the strength and the diameter of the vortex could be varied by controlling the speed of the small displacement of the plate in the upstream direction.

The configuration of the water surface around the vortex axis was measured by a wave gauge installed downstream. The vortex motion as well as the upwelling motion induced from the channel bottom were made visible by means of different dyes. Both the side view as well as the overhead view (reflected by a mirror set above the basin) were photographed every second. The experiments were performed with two types of bottom conditions : first a smooth surface, second a rough bottom made of pasted sand particles of 1.7 mm, and 2.2 mm diameter. The wave gauge was installed 0.5 m downstream from the vortex generation point. The turbulent level caused by the gauge was estimated to be negligible.

The experimental results are presented in Table 2, where H is the water depth and Γ the strength of the vortex. The vortex strength was estimated using the Rankine's combined vortex model, i.e.,

$$\Gamma = 2 \pi a \sqrt{g \cdot \Delta h} \tag{12}$$

where Δh is the maximum water surface depression at the vortex center and a denotes the radius of the contour line of $\Delta h/2$. Both the values of Δh and a were measured from the vertical section of the vortex recorded by the wave gauge. The value of the tangantial velocity, V_{mo}, was calculated by the relation $V_{mo} = \Gamma/2\pi a$. The value of the upwelling velocity, W_{mo}, was obtained from the displacement speed of the upwelling flow, estimated from a sequence of

Vortex No.	①	②	③	④	⑤	⑥	⑦	⑧	⑨	⑩	⑪	⑫	⑬	⑭	⑮	⑯	⑰	⑱	⑲	⑳	㉑	㉒
RADIUS R_0 (cm)	2.08	3.60	3.02	3.82	3.96	1.62	2.28	2.74	3.36	3.19	3.28	2.31	2.58	3.67	4.21	1.91	3.28	3.36	4.18	3.12	2.61	3.57
SURFACE CONCAVITY $\Delta\lambda$ (cm)	0.63	0.35	1.18	0.88	1.88	0.42	0.70	0.56	1.16	1.93	0.25	0.93	2.46	1.25	1.03	0.82	0.31	0.31	0.39	1.18	2.30	2.38
VORTEX STRENGTH Γ (cm^2/sec)	325	419	645	705	1068	237	375	498	712	871	319	438	796	807	840	340	359	368	513	667	778	1063
UPWELLING VELOCITY W_{mo} (cm/sec)	2.0	2.5	2.8	2.8	4.6	3.3	3.8	3.3	5.0	5.0	3.0	3.9	4.2	5.0	5.0	5.4	3.5	5.0	3.2	3.0	2.0	4.0
TANGENTIAL VELOCITY V_{mo} (cm/sec)	40	18	34	29	42	20	19	28	33	43	15	30	49	35	31	28	17	17	19	34	47	48
SWIRLING RATIO G	20.0	7.4	11.0	10.0	9.2	6.1	5.0	8.5	6.7	8.6	5.1	7.6	11.0	7.0	6.3	5.2	4.9	3.5	6.1	11.0	23.0	12.0
DEPTH H (cm)	20					20					30					30						
BOTTOM	smooth					rough					smooth					rough						

TABLE 2. Experimental results

photographs. The swirling ratio G was estimated from the values obtained for V_{mo} and W_{mo}. It is expected that all experimental cases in the study are in the range of the ring-like upwelling condition, because the swirling ratio is always larger than one.

Fig. 11 shows the successive pictures of the upwelling flow mapped from a series of photographs with a time interval of one second. The upward volume flux of the upwelling motion was estimated from similar figures obtained for each of the experiments listed in Table 2. The results are represented in Fig. 12 as a function of the vortex strength. The upward volume flux caused by the vortices in the Naruto strait was estimated from the experimental results on the basis of an assumption of Froude similarity. In the experiments, a flux of 500 cm^3/sec was observed to be caused by a vortex of 3 cm diameter, 0.8 cm water surface depression and 500 cm^2/sec strength, for a water depth of 20 cm. In the Naruto strait, under flood tide condition, the vortices are about 6 m in diameter, and have a 1.6 m surface depression and a strength of 150 m^2/sec for a water depth of 40 m. Therefore, the upward volume flux due to a single vortex is estimated to be about 280 m^3/sec.

In the Naruto strait, about a dozen of vortices of this importance
are observed as shown in Photo. 1. As a result, these vortices are
expected to cause a total upward volume flux of about 3,000 m^3/sec.
The total volume flux of the tidal current is approximately
100,000 m^3/sec at the maximum stage of a flood tide. Therefore,
in the Naruto strait the vertical mixing due to the upwelling flow
associated with the vortices is estimated to be a few percents of
the total volume flux of the tidal current passing through the
strait.

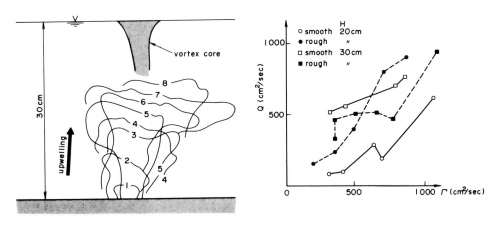

Fig. 11. Successive pictures of   Fig. 12. Upwelling volume flux
the upwelling flow

PREFERRED MODE OF SURFACE WATER JETS IN THE NARUTO STRAIT

    Regarding the main hydrodynamic feature of the tidal current,
namely its surface jet behaviour, the so-called "preferred modes"
of the jet is an interesting subject to discuss. In connection to
this problem, it has been recently reported that the impingement
of separated shear layers on solid boundaries generates a feedback mechanism which sustains oscillations at selected frequencies
within the band of frequencies at which the shear layer is unstable. Owing to the instability of the shear layer, small perturbations in the flow development zone are amplified to form vortices
of coherent structure, as they travel downstream. The induced force
associated with the impingement of these vortices on the solid
boundary produces an upstream influence which modulates the sensitive region of the shear layer near separation and then gives rise
to new vorticity perturbations. Ho and Nosseir (1981) found that a

strong resonance develops in the flow if the solid boundary is
located about six times an   exit diameter downstream from a round
nozzle and if the Mach number is above approximately 0.7. Laufer
and Monkewitz (1980) have analyzed a free jet and found that the
amplitude of the unstable oscillations are strongly modulated at
Strouhal number St=FD/U=0.31, where F is the frequency, D the
nozzle diameter, and U the exit velocity. Furthermore, from experimental studies of air jets, it has been reported that the Strouhal
number in the preferred mode would be in the range of 0.25<St<0.5
under subsonic conditions. In the case of surface water jets, such
critical values of the Strouhal number have not been considered
yet. Especially, measurements of the preferred mode of water jets
in nature have not been reported before. The author tries to
estimate the preferred mode in the Naruto strait below.

From existing studies of air jets, it can be inferred that
vortices coalescing in the shear layers in water surface jets will
be generated under certain flow conditions, and that the feedback
mechanism will be stimulated by the intermittent passage of vortices at the end of the potential core, instead of by the impingement on the solid boundaries located in the air jets. Further, it
is surmised that while the shear layers in the air jets are desta-
bilized due to the perturbation of the pressure, the instability in
the water jets shall be induced by a surface undulation propagating
upstream. Therefore, the feedback mechanism should exist only in
subcritical flow.

On the basis of the above considerations, the vortices observed
in the Naruto strait, such as shown in Photo 1 and Photo 2, can
be regarded as the results of amplified waves sustained at the
preferred frequencies. Introducing a nozzle width L as a characteristic length of the flow, an exit velocity V as a characteristic
velocity, and a reciprocal of the passage interval of the vortices
at the end of the potential core, $1/\tau_o$, as a characteristic
frequency, the Strouhal number is represented by

$$St = L/V\tau_o,\tag{13}$$

where τ_o corresponds approximately to the value of $\tau = b/c$ at the
downstream end indicated in Table 1. An exact determination of the
nozzle width of the surface jets in the Naruto strait is rather
difficult. But from Photo 1 and the known characteristics of the
region where the jets are observed, the location of the jet nozzle

in the strait can be assumed as shown in Fig. 13, and its width
estimated at about 120 meters. The mean water depth in the assumed
nozzle section is approximately 100 meters, and the cross sectional
area of the exit flow is estimated to be about 12,000 m^2. For the
section connecting the head of Oge island and Tozaki cape, the
average velocities of the jets are shown in Table 1 and the cross
sectional area is about 32,000 m^2. Hence, using the principle of
continuity, the exit velocity V in the assumed nozzle becomes

$$V = 2.7 \ U_0. \tag{14}$$

By substituting eq.(14) into eq.(13), the Strouhal number becomes

$$St = 44.4/U_0 \ \tau_o \tag{15}$$

With the above relation, the Strouhal number was calculated for the
vortices observed along the shear layer located to the right of
the downstream direction. The results are presented in Table 3.
The Strouhal numbers at the end of the potential core in the strait
are of order of 0.2 to 0.5. These values are approximately equal
to those of the round nozzle impingement air jets, as well as of
the free air jets.

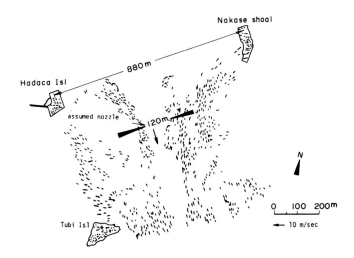

Fig. 13. Assumed location of nozzle in Naruto strait

128

Date	Apr. 1st, 1977	May. 8th 1977	Jul. 2nd, 1977	Feb. 24th, 1978
St	0.22	0.25	0.51	0.30

TABLE 3. Estimated Strouhal number of preferred mode in Naruto jet

BEHAVIOUR OF THE JET IN THE ZONE OF DEVELOPED FLOW

Photo 5. The thermal imagery of the southward current.

Photo 5 is a thermal image of the maximum southward tidal current. The photograph was taken from an altitude of 4,000 meters at 12:00 a.m. on Aug. 23rd, 1979. Darker tone in the photograph corresponds to higher temperature. In Photo 5 one can see that the warmer tidal jet penetrating into the Harrima channel takes the shape of a puff of lower temperature in the zone of developed flow. In the case of a stratified surface jet penetrating into a large stagnant environment of infinite depth, the water surface temperature in the zone of developed flow shall gradually decrease in the downstream direction. In the Naruto strait, however, the jet reaches the sea bottom and, as discussed earlier, strong vertical and horizontal mixing takes place, producing the puff of lower temperature. The shape of the jet in the zone of developed flow is influenced by various parameters such as the bottom slope, the bottom roughness, the aspect ratio of the exit, the fluid viscosity and the density difference. To investigate these effects, hydraulic model tests were performed in a 2.5 m x 1.8 m x 0.3 m basin. First,

the effects of the bottom friction and of the exit aspect ratio
were studied. In the experiment, the horizontal shapes of the jet
at a fixed time interval were recorded in a series of photographs
and the changes of the jet width with time were estimated. The
result is presented in Fig. 14, using a non-dimensional parameter
$\mu = fb_o/8h_o$ (f = friction factor, b_o = half width of the jet;
h_o = water depth at the exit). Dotted lines in the figure are
Ozosoy and Unluata's analytical results (1982). Both the experimental and analytical results indicate that the jet width increases
nonlinearly with the distance from the exit in the presence of
bottom friction, whereas the relationship is linear in a non-vis-
cous fluid. In Fig. 14 one can see that the experimental and
analytical results show good agreement in the region near the
exit.

Second, the effect of the density stratification was investigated,
using salinity differences to stratify the fluid. In a buoyant
surface jet, the depth of the jet increases first in the inertia

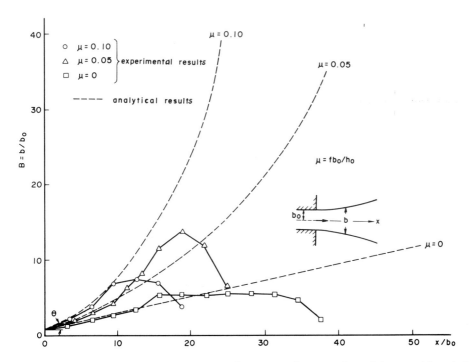

Fig. 14. Effects of bottom roughness and aspect ratio on the jet
width in a homogeneous fluid.

dominated region and then decreases due to the buoyancy. Therefore, the jet width in the zone of developed flow should depend on whether the maximum depth of the surface flow is larger than the depth of the receiving region or not. Stolzenbach and Harleman (1971) suggest that the maximum depth of a buoyant surface jet is given by $h_{max} = F_o \sqrt{h_o b_o}/2$, where F_o is the internal Froude number defined by

$$F_o = u_o / \sqrt{g \frac{\Delta\rho}{\rho} h_o},$$ (16)

and where g is the gravitational acceleration, u_o the exit velocity, ρ the density of the receiving region, and $\Delta\rho$ the density difference between jet and receiving water. Fig. 15 (a) shows the variation of the jet width with time for $F_o = 3.42$ and h_{max} smaller than the receiving water depth. Fig. 15 (b) is the result obtained in the same conditions except for h_{max} being larger than the receiving water depth. By comparing these figures, one can see that the internal surface increases the jet width more effectively than the bottom in a stratified fluid. In the case shown in Photo. 5, although the stratification may be rather weak, the flow conditions would qualitatively correspond to those of Fig. 15 (a). The same can be said of a flow pattern obtained through analysis of Landsat data for the Naruto strait under the condition of maximum northward tidal current at 9:48 a.m. on Aug. 1st, 1976.

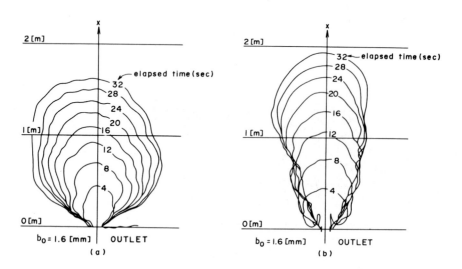

Fig. 15. Spreading of a buoyant jet in stratified water

SUMMARY AND CONCLUDING REMARKS

The tidal currents in the Naruto strait present the typical
appearance of surface water jets, because of the particular configuration of the strait. In particular, the vortices along the shear
layers in the flow development zone show dynamic features often
observed in the coastal waters of Japan. But the fluid dynamic
characteristics of the vortices, such as circulation, vertical
profile, dynamic structure, dominant mode and others, have not been
quantitatively measured, perhaps because of lack of suitable
surveying methods. However, recent remote sensing techniques seem to
provide a way to overcome these difficulties.

In this paper the author describes the fluid dynamic characteristics of the tidal jets in the Naruto strait with the help of several remote sensing techniques including aerial photography,
infrared photography, colour sonor and Landsat data. The water
surface depression at the vortex centre is measured by the method
of synchronous aerial photography.

From these results and the application of the Kameron effect, the
circulation of the vortices is estimated to be about 100 m^2/sec in
the Naruto strait. The application of the colour sonor to observe
the vertical profiles of the vortices in nature seems noteworthy.
Its results indicate that the vortices along the shear layers in
the Naruto stait have coherent structures resembling columns
connecting the water surface to the sea bottom. The results also
indicate that these vortices generate secondary upward flows. These
phenomena have been verified in hydraulic model tests. The experimental results suggest that the vertical mixing due to the upward
flow associated with the vortices is a few percent of the total
volume flux of the tidal current passing through the strait. The
observation of the structure of the potential core, by infrared
photographs of the water surface, reveals the existence of many
small vortices in the core region. Some of these vortices are
distributed in a series of "rollers" as shown in Photo. 3.

Other interesting results are those regarding the so-called
"preferred mode" of the unstable shear layer. From an analysis of
the data of the water surface jets in the Naruto strait, it is
concluded that the dynamic vortices such as shown in Photo. 1 and
Photo. 2 are generated along the shear layers due to its instability when the Strouhal number at the end of the potential core is
about 0.2 to 0.5. These values of the Strouhal number in the

preferred mode are approximately equal to the ones reported in investigations of impingement air jets.

The characteristics of the tidal jets in the zone of developed flow are discussed in a rather qualitative way with application of infared photography and Landsat data as well as hydraulic model tests. Details of the application of the Landsat data have been reported in other publications by the author and his colleagues.

REFERENCES

Ho, C. and Nosseir, N.S., 1981. Dynamics of an impingement jet, part 1. The feedback phenomenon. J. Fluid Mech., 105, 119-142.
Laufer, J. and Monkewitz, P., 1980. On turbulent jet flows; A new prospective AIAA paper 80-0962, Hortford, Conn.
Maruyasu, T., Onishi, S. and Nishimura, T. 1981. Study on tidal vortex in the Naruto strait through remote sensing. Bull. of Remote sensing lab. Rem. Sen. series n°1, Scie. univ. of Tokyo.
Onishi, S. and Nishimura, T., 1980. Study on vortex current in strait with remote sensing.
The 17th Int. Coast. Eng. Conf. ASCE, pp. 2655-2670.
Ozsoy, E. and Unluata, U., 1982. Ebb-tidal flow characteristics near inlets. Estuaring, Coast. and Shelf Sci. Vol. 14, pp. 251-262.
Schlichting, H., 1968. Boundary layer theory, Sixth Edition, McGraw-Hill, 214, pp.
Scorer, R.S., 1978. Environmental aerodynamics, Ellis Horwood Publisher, p. 229.
Stolzenbach, K. and Harleman, 1971. An analytical and experimental investigation of surface discharges of heated water, M.I.T. Hydro. Lab. Tech. Rep., n°135.

SURFACE TEMPERATURE AND CURRENT VECTORS IN THE SEA OF JAPAN FROM
NOAA-7/AVHRR DATA

T. SUGIMURA[1], S. TANAKA[1] and Y. HATAKEYAMA[2]

[1]Remote Sensing Technology Center of Japan, 7-15-17, Roppongi,
Minato-ku, Tokyo 106 (Japan)

[2]Asia Air Survey Co., Ltd., 13, Nurumizu, Atsugi-city, Kanagawa-
ken 243 (Japan)

ABSTRACT

This paper describes hypotheses about the circulation in the Sea
of Japan, and gives proof that these hypotheses do not contradict
the sea surface temperature pattern or the current vectors obtained by NOAA/AVHRR data, and sea surface temperature recorded by
ships. The hypotheses are as follows : The water of the Sea of
Japan moves in a counter-clockwise direction in general. The
Tsushima Current flows northward along the coastline of the
Japanese Islands as a surface current. On the other hand, the
Liman Current flows southward along the shoreline of Siberia as a
surface current. In addition, it is postulated that a deep current
rotates along the edge of the Nippon Basin. This deep-water rises
up to the sea surface in the winter season. Especially, it appears
on the surface offshore of Vladivostok, probably because of a
reaction to the sea mountain front off Vladivostok. Another current
may be a vertical circulation existing between Japan and Siberia
and caused by a seasonally prevailling wind.

INTRODUCTION

Current conditions in the Sea of Japan have been investigated by
means of tracking drift bottles, or measuring the sea surface temperature since the late 19th century. Recently, a new method has
been developed to determine the current field from satellite data.
The method has been exploited to obtain Kuroshio Current vectors
from two adjacent NOAA/AVHRR data (with a 12 hour difference
between them). In the experiment reported here the authors have
investigated the current field in the Sea of Japan, using NOAA
data, actually using two temporal NOAA-7/AVHRR imageries taken on
27 October 1982. Current vectors can be obtained by comparing the
corresponding sea marks on the imageries. From the study of these
imageries, the authors obtained a good indication of the current
pattern in the Sea of Japan, comparing satisfactorily with observations of seagoing vessels.

134

HORIZONTAL CIRCULATION

Currents in the Sea of Japan were investigated scientifically by exploring ships at the very beginning of these studies more than 100 years ago. A current chart of Japanese waters was presented by Shrenk in 1873 as shown in Fig.1. This chart was hypothetic and was based on water temperatures observed by merchant and war ships. This is not different from the method now being used to survey ocean currents from ocean temperatures.

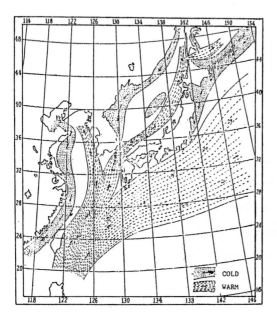

Fig. 1. Current chart around Japan (after Shrenk 1873).

This chart is interesting because of the fact that the location of cold and warm waters is fairly similar to that shown on the surface temperature map obtained by satellites (see Fig. 3C). After the beginnings of such studies, various charts of the ocean current were compiled. Among them, the chart presented by Dr. Wada in 1916 (as shown in Fig. 2) is related to the authors' theory on the rotating circulations in the Sea of Japan. Three counter-clockwise circulations can be recognized and the inner ring appears to circu-late along the equi-contour line of the basin of the Sea of Japan at a depth of 3,000 m. This chart was compiled by tracking drift bottles.

Fig. 2. Current chart around Japan (after Wada, 1916).

Fig. 3 shows surface temperatures of the Sea of Japan compiled
from IR (channel 4) data of NOAA-7/AVHRR obtained on October 27,
1982. The pictures at the left and the right present the night and
the day patterns, respectively. If we assume a law in which the
stream lines are generally parallel to the isotherms, ocean
currents ruling the Sea of Japan can be estimated. Two major
currents can be seen in the Sea of Japan. One is the so-called
Tsushima Current, a warm current flowing through the Tsushima
Strait, branching of the Kuroshio. The other is the so-called Liman
Current, a cold current flowing along the coast of Siberia from
the north-eastern sea area. The Tsushima Current flows north along
the shoreline of the Japanese islands with a velocity of about 0.2
to 0.6 knots. Referring to the current map compiled by the Maritime
Safety Agency of Japan (shown in Fig. 4), current velocities are
relatively high (1.0 to 1.9 knots). However, the distribution of
current directions is not uniform but, rather, complicated. This is
probably due to the fact that currents in each finite sea area are
affected by local winds. In order to define prevailing current
movement, the temperature maps of the sea surface and the water
below the surface become rather important. NOAA imageries in (A)

136

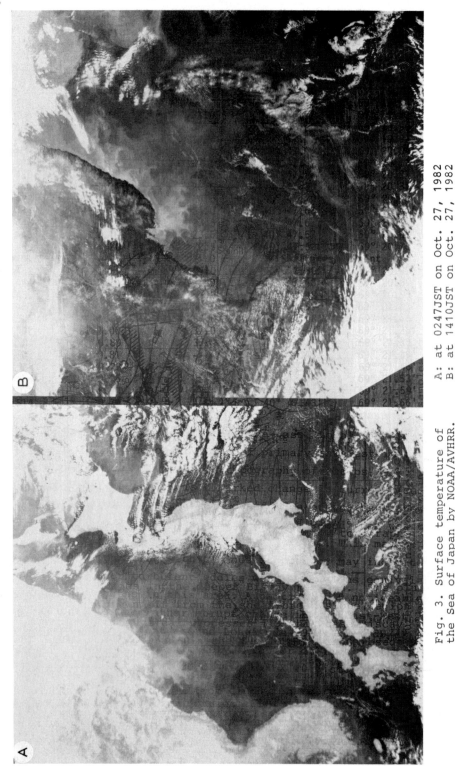

Fig. 3. Surface temperature of
the Sea of Japan by NOAA/AVHRR.

A: at 0247JST on Oct. 27, 1982
B: at 1410JST on Oct. 27, 1982

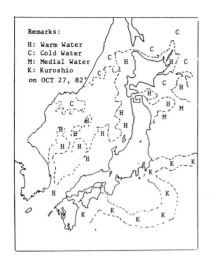

Fig. 3C. Warm and cold water around Japan by NOAA/AVHRR.

and (B) of Fig. 3 show the surface temperature. Temperature can be calibrated by using surface temperature maps compiled by the Maritime Safety Agency. True temperature of the sea surface and the water at 100 m depth are shown in Fig. 5.

The prevailing current chart of the Sea of Japan was estimated as in Fig. 6. This chart shows only the current directions but not their velocities. Features of the horizontal currents of the Sea of Japan can be summarized as follows :

(1) There are two major currents, namely the Tsushima Current and the Liman Current. The Tsushima Current flows from south-west to north-east along the Japanese Islands. On the other hand, the Liman Current flows from north-east to south-west along the Siberian shoreline.

(2) The shape of the front formed where the two currents meet is complicated.

It is meaningful to discuss, just here, the sources of energy which induce the current movements in the Sea of Japan. Two major sources can be considered. The first one is the Tsushima Current which constantly flows along the shoreline of the Japanese Islands. The flux has been estimated in the range of 2 to 8×10^6 m^3/sec. A value observed in September of 1964 was reported as 4.2×10^6 m^3/sec at Tsushima Strait. This flux varies with the seasons. The second source of energy is the wind. Wind

138

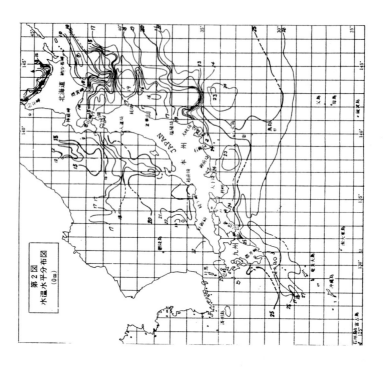

Fig. 5A. Horizontal distribution of water temperature (0 m) from Oct. 14 to Nov. 4 of 1982 (by Maritime Safety Agency of Japan).

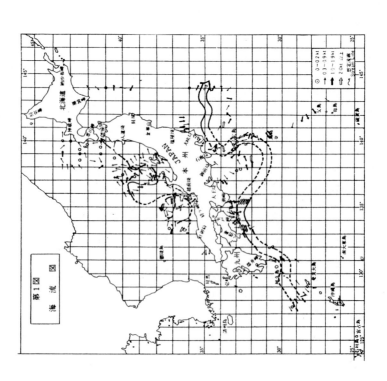

Fig. 4. Current chart from Oct. 14 to Nov. 4 of 1982 (by Maritime Safety Agency of Japan).

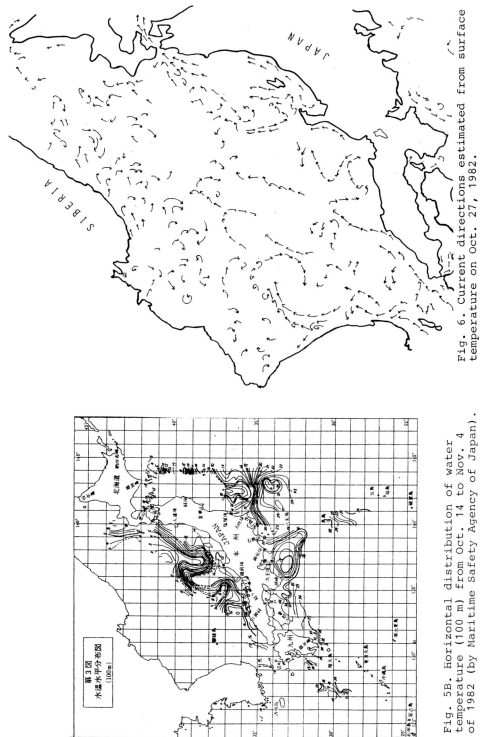

Fig. 6. Current directions estimated from surface temperature on Oct. 27, 1982.

Fig. 5B. Horizontal distribution of water temperature (100 m) from Oct. 14 to Nov. 4 of 1982 (by Maritime Safety Agency of Japan).

induces the skin currents. Prevailing wind directions are from
Siberia to the Japanese Islands.

CURRENT VECTORS FROM SATELLITES

Using serial data taken from two different time periods (as shown
in Fig. 3A and 3B), current vectors in the Sea of Japan were
obtained as shown in Fig. 7.

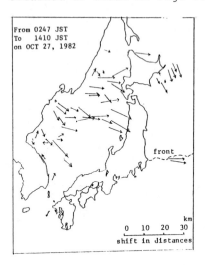

Fig. 7. Current vectors around Japan by NOAA/AVHRR

The measuring accuracy was 1.5 km in the east to west direction and
0.5 km in the north to south direction. These values were verified
at control points shown in Table 1, where 1 km corresponds to
0.05 knots because of the time difference of 11h23m in the two sets
of data.

TABLE 1

Measurement accuracy of seamark shift.

No.	Check Point		Residuals		Land Marks
	ϕ	λ	X(Eastward)	Y(Northward)	
1	46°33'41"	138°20'58"	−1.8 km	0.7 km	Tip
2	41°22' 8"	129°48'54"	1.4 km	−0.5 km	Bay
3	35°42'26"	140°52' 0"	−0.5 km	−0.2 km	Tip
4	36° 4'52"	129°33'23"	0.9 km	−0.3 km	Tip
			$\sigma_{x,n-1}$=1.5km	$\sigma_{y,n-1}$=0.5km	

From the authors' traditional concept of the currents in the Sea
of Japan, the vectors should be from south-west to north-east along
the western side of the Japanese Islands, and from north-east to
south-west along the eastern side of Siberia as shown in Fig. 1.
However, the directions of the vectors obtained from satellites
are almost uniform and from north-west to south-east. The velocity
is less on the Siberian side than on the Japanese side. For
example, 0.23 knots is the velocity on the Siberian side, but
0.43 knots is the velocity in the central portion, and 0.52 knots
is the velocity about 200 km offshore Japan.

VERTICAL CIRCULATION

The force inducing the current field seen in Fig. 7 is the wind
stress. For example, taking a vertical view at the cross section
line A-A in Fig. 8, one obtains the vertical circulation given in
Fig. 9. On the water surface, current velocities are shown as small
on the Siberian side and large in the central sea area. This
current may be the so-called skin current generated by the wind.

The underwater circulation system cannot be measured, but might
be as shown in Fig. 9. Surface water on the Siberian side is
carried towards the central area of the Sea of Japan. Therefore,
a new water mass in this area should be supplied by some mechanisms.
It is natural to think that this new water comes from the deep sea.

This assumption is not inconsistent with the surface temperature
distribution in which a lower temperature appears on the Siberian
side but grows higher as one gets further away from the shoreline
as in Fig. 3. The authors believe the vertical profile of the
horizontal current to be as shown in Fig. 10.

First, the highest reverse velocity might exist at a depth of
about 1,000 m, because the lowest temperature zone is located at
this point as shown in Fig. 11. Next, two depth levels have zero
velocity in this vertical section, namely at a point between the
surface and 1,000 m, and at the bottom.

Maximum speed of deep water currents towards Siberia might be
0.02 knots or less at the Nippon Basin. This is a result of the
vertical temperature profile of Fig. 10. The deep water current
velocity reaches merely about 1/10 of the prevailing skin current
velocity. Although this value is derived under the condition that
the skin current velocity is reduced in linear form, actual velocity reduction is higher than temperature reduction.

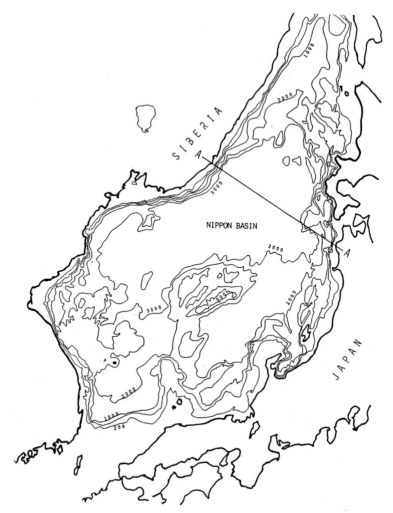

Fig. 8. Bottom topography of the Sea of Japan.

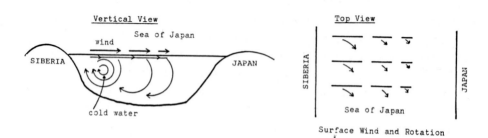

Fig. 9. Vertical circulation caused
by wind stress.

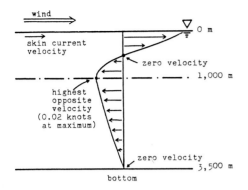

Fig. 10. Vertical profile of the horizontal current.

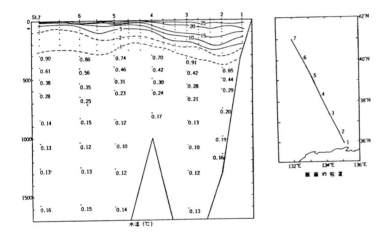

Fig. 11. Vertical temperature profile observed in Aug. 1964 (after Moriyasu 1972).

HORIZONTAL CIRCULATION OF DEEP WATER

When energy is introduced to enclosed water bodies in the northern hemisphere, water begins to circulate counter-clockwise. Main energy sources are supplied from the Tsushima Current through the Tsushima Strait, and by seasonally prevailing winds.

The mean velocity of the Tsushima Current was measured as being about 0.4 knots along the shoreline of the Japanese Islands. The mean velocity of the skin current induced by prevailing winds might also be 0.2 to 0.4 knots.

These velocities generate a secondary velocity by the Coriolis

effect. Along the circular edge of the Nippon Basin, velocity of
surface currents rotating counter-clockwise can be equal to or less
than 0.2 to 0.4 knots.

The vertical distribution of this current can be predicted as
shown in Fig. 12. The authors doubt that the velocity is actually
as large. It seems that the velocity should be 1/10 of the tangential component of the current vectors as in Fig. 7. In the deep
sea, current velocity slows down in accordance with the depth as
shown in Fig. 12.

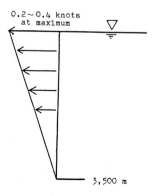

Fig. 12. Predicted current distribution in deep sea.

PROOF OF DEEP WATER CIRCULATION

The cold pattern offshore of Vladivostok looks like turbid water
discharged from rivers. The authors thought so at first glance.
However, this pattern is not a discharge from rivers, but more
likely cold water upwelling from the deep sea. The scale of this
cold water area is large and its diameter is beyond 400 km. Now,
the authors believe that this is proof of deep water circulation.

If one assumes that the circulation illustrated in Fig. 13 is
a deep water circulation, this upwelling phenomenon becomes
possible.

The velocity of surface water circulation is not high, 0.2 to 0.4
knots at most. Therefore, the deep water current is reduced to a
small value in accordance with the water depth. Furthemore, deep
water currents may cease or reverse, with the season.

Upwelling phenomena offshore of Vladivostok occur clearly in
October. The conditions under which upwelling can be accelerated

145

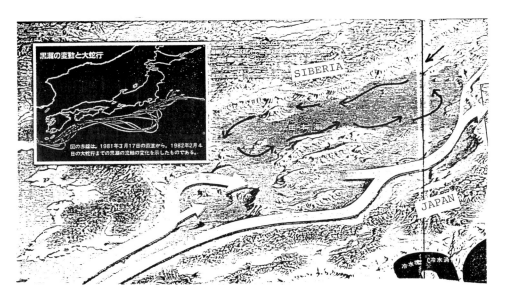

Fig. 13. Deep water circulation around Nippon Basin (after "NEWTON" 1983).

are :

(1) When the deep water current is strong.

(2) When the surface water density is heavy.

We have an idea of the surface temperature distribution of the Sea of Japan, also from conventional observation carried out by the Japanese Navy about 60 years ago : a relatively cold patch is shown in the winter chart, but it is not distinguishable in the summer season. These two sets of temperature charts suggest to us that upwelling has occurred in the winter season since long ago. Presumably, the surface water is frozen over the winter season and it changes to heavy water. Under this condition, upwelling can occur very easily.

The authors also investigated other seasons of NOAA/AVHRR imageries, September 16, 1981, and March 3, 1982. The temperature differences between the center of the upwelling and ambient waters were found to be of 9.3°C, 3.0°C, and 1.0°C in October, September, and March respectively. This shows that the strongest upwelling occurs in October, followed by September, and then March.

CONCLUSION

From the observations and concepts on the current phenomena of the Sea of Japan, the following points may be suggested :

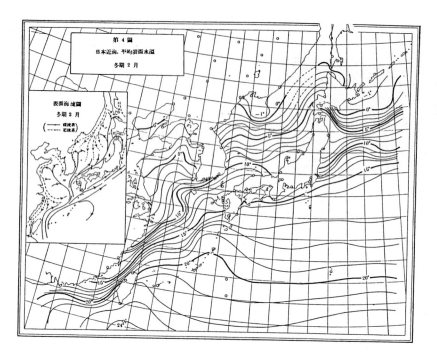

Fig. 14A. Surface water temperature around
Japan observed in Feb.(after Suda in 1930).

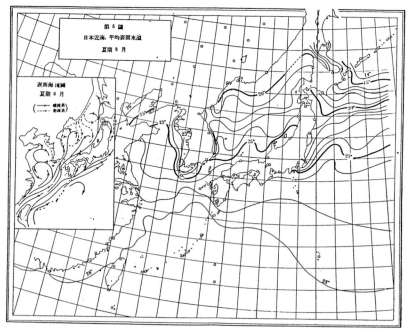

Fig. 14B. Surface water temperature around
Japan observed in Aug.(after Suda in 1930).

(1) A detailed pattern of the stream lines can be deduced from temperature distribution given by NOAA data. The current distribution is a result of the interpretation of the shape of the frontal line. The current distribution is never simple.

(2) Vertical circulation between Siberia and Japan can be postulated, because of the uniform directions of the skin currents on the whole sea surface. Actual deep water velocities remain subject for further studies.

(3) A horizontal deep water circulation may exist, because there is very strong upwelling offshore of Vladivostok. This phenomenon cannot be explained without such a circulation.

ACKNOWLEDGEMENTS

The authors express their thanks to Prof. S. Onishi for his general suggestions on their report. Also, they thank Mr. T. Jinguji and M. Murata of Hosei University for their assistance in supplying figures for the report.

REFERENCES

Suda, K., 1930. Ocean Physics. Iwanami.
Suda, K., 1933. Ocean Science. Kokonshoin.
Moriyasu, S., 1972. Ocean Physics in the Sea of Japan. Ocean Science, March.
Government Report, 1982. Kaiyo Sokuho (Quick review of ocean condition). 4 Nov. 1982.
Tanaka, S. et al., 1982. Compilation of the Kuroshio Current vector map from NOAA-6/AVHRR data and consideration of oceanic eddies and the short period fluctuation of the Kuroshio. J. of RSSJ, Vol.2, No.3.
Sugimura, T. et al., 1983. Maritime condition of the Sea of Japan in Autumn 1982. Falls Convention of OSJ, Apr.
Graphic Science Magazine NEWTON, 1983. Kyoikusha, Apr.

STUDY OF MESOSCALE PROCESSES IN THE SHELF ZONE OF THE BLACK SEA
USING REMOTE TECHNIQUES

R.V. OZMIDOV and V.I. ZATZ

Institute of Oceanology P.P. Shirshov, Moscow

Remote techniques offer several advantages over contact techniques for the study of diffusion and other mesoscale hydrodynamic
processes in the sea. First, remote techniques provide an instantaneous survey of a given spatial domain which is impossible to
obtain by contact techniques. Second, remote techniques can also
provide a series of instantaneous surveys at given time intervals;
such a series can be used to investigate the time variability of
a phenomenon over a wide range of scales. A third advantage is the
possibility to investigate the diffusion of a passive scalar and
other oceanic processes on scales of the order of one to ten metres,
while avoiding the rather significant distortions that can be produced by a moving vessel or device.

Airphotometric studies of diffusion in the sea started in the
sixties, when Japanese scientists (Ito, 1964; Ito et al., 1966);
Fucuda et al., 1964; Nanniti, 1964) performed air photography of
dye patches in a coastal zone. These papers and subsequent ones
(Ichiye, 1965; Pritchard et al., 1966; Ichiye, 1967; Assaf et al.,
1971) concerned the study of the patterns of the patches, arising
from streaks of different scales, and of the influence of hydrometeorological conditions on the patches. Djuric and Leribaux (1974)
used sinked cans in their experiments, and air photography permit-
ted to investigate the dispersion of a cloud of particles as a
function of time. Ichiye and Plutchak (1966), and Ichiye (1967)
used air photography of patches and jets of dye to investigate
the spectral structure of the dye concentration field. Further
studies of this phenomenon were performed by Schott et al. (1978)
and James and Burgess (1970). In the latter paper, air photography
data were compared with data of contact measurements of rhodamine
concentration in a sewage flow.

Our studies of diffusion in the shelf zone of the Black Sea using
aerial photography started in 1968. The aim was to find relation-

ships between the size of the patches, their area, the exchange
coefficients and the growth rate as a function of hydrometeorological conditions. To that aim, complex investigations were performed using remote methods as well as contact hydrological and fluo-
rimetric measurements.

For airphotometric methods we used photocameras whose resolution
in the center of the patch was 40 lines per millimetre, and at the
boundary up to 20 lines per millimetre. The shooting film was made
of various photomaterials including spectral film; a number of
different light filters was used. The frame size was equal to
30 x 30 cm or 18 x 18 cm. We performed a systematic study of the
dependence of the quality of the films on the illumination of the
sea, the sea surface state, the characteristics of the cameras and
the methods of shooting. The deviations of the main axis of the
camera did not exceed 2°; the accuracy of determination of the
height of the flight by radioaltimeter was 1,5 m. The scale of
the films varied from $1 \div 2500$ to $1 \div 7500$. Since the estimated
errors on the aerial photographs increase towards the periphery
of the pictures, the processing of the image was often limited to
a central zone wherein the error did not exceed some threshold
value.

In the processing of air photographs of dye patches, it is important to relate the actual dye concentration in the upper layer of
the sea to the integrated concentration over some area of the sea
determined by photographical method. To determine the degree of
darkness of a negative we used microphotometers and densitometers,
in particular, the "OPTRONICS - PDR - 1" system in which a video-
image is transformed into digital form and inputed into a computer.
The number of discretization levels for density of darkness was
equal to IO. The calibration of the photographs was carried out
using simultaneous contact measurement of dye by fluorimeters.
For a similar calibration, Schott (1978) suggested a power relation
between the observed concentration and the photodensity of the
negative. Unlike Schott, we used in our calibration not only towed
fluorimeter data but also data of soundings.

Aerial photographs of dye patches taken in 1981-1982 over coastal
areas off the Crimea and near Sukhumi were used to study the variability of patch sizes over a time span of 10^2 to 10^5 sec following
the release of the dye as a point source. The patch area, defined
using some threshold isoline of concentration, increases with time

according to a power law whose exponent falls in the range 1,11 to 1,60. When the patch area approaches its maximum value, the rate of its growth slows down. In a number of cases, we observed several stages of area growth with different exponents. The equivalent radius of the patches, denoted r, also increases exponentially with time. A linear dependence is often observed immediately after the release of the dye and persists over diffusion time of 20-40 min. An exponent equal to 3/2 (which corresponds to the inertial range of turbulence) is rather seldom observed, and usually during storm winds. An exponent of 1/2 (regime of Brownian motion) is often observed when the patch area approaches its maximum value. Consequently, over diffusion times ranging from 10^2 to 10^5 sec, a succession of several diffusion regimes is observed (Fig. 1).

The degree of anisotropy α of the patches, defined as the ratio of the length of the maximum axis of the patch to its mean radius, is also a function of time. During the early stages of diffusion (30-60 min after release), the patch shape is very close to a circle. At later stages, the patch becomes elongated. The value of α varies usually from 1 to 15 (Fig. 2). Using the data on maximum radii of the patches and on diffusion time, we can also define a diffusion velocity P (by the Joseph-Sendner method). The calculated values of P are in the range of 0,31-2,1 cm/sec and they decrease with time (Fig. 3).

The values of the coefficients of horizontal turbulent exchange were determined using air photography data for diffusing floats loaded with packages of dye. The photographs were taken every 4-5 minutes during several hours. The scale of the phenomenon (distance between floats) changed from 10 to 474 m. The values of the exchange coefficients were in the range $10^2 - 10^5$ cm^2/sec (Fig. 4).

The data of air photography of dye patches suggest that shear effects in the velocity field of nearcoastal currents are important. The values of the shear were calculated using current velocity data obtained from buoy stations and by means of drifting floats. These values were in the range $(0,1-14,5).10^{-5}$ sec^{-1}, depending mostly on the distance between instruments (Fig. 5). For the Caucasus coastal region this dependence is stronger because the shore is more straight and the flow structure is similar to that of a turbulent boundary layer. The vertical velocity shear in the upper 100 m layer was determined using the data of current meters

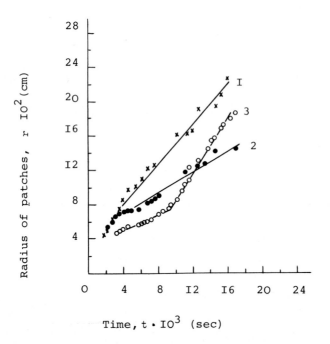

Fig. 1. Time variation of the mean radius of a diffusing patch, denoted r, during the course of three experiments.

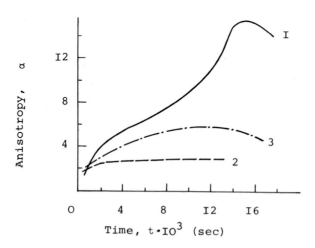

Fig. 2. Time variation of the anisotropy parameter, α, for three experiments.

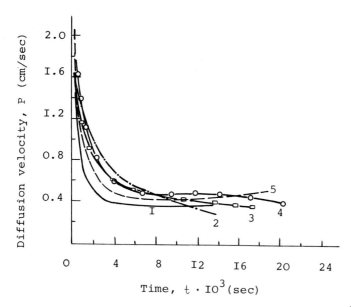

Fig. 3. Time variation of the diffusion velocity, $P = \dfrac{r_{max}}{2t_{max}}$, for several experiments.

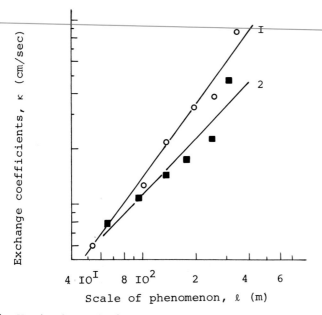

Fig. 4. Variation of the coefficient of horizontal diffusion $\kappa = \dfrac{(\Delta \ell)^2}{2\Delta t}$, with the scale ℓ of the phenomenon estimated from air photographs of floats. Experiment n°1 took place in the Caucasus coastal region, experiment n°2 in the Crimea coastal region.

at buoy stations. The vertical shear can be as high as 10^{-1} sec^{-1}; the mean value is 10^{-2} sec^{-1} which exceeds the mean values usually observed in the open ocean.

The elongation of the dye patches usually occurs in the direction of the current in the absence of wind or when the wind blows in the direction of the current. If the directions of the current and of the wind are different, the axis of the patch can be aligned with yet another direction (Fig. 6). The elongation of a patch can be influenced by the vertical velocity shear when deeper layers of the patch "lag behind" nearsurface layers. An example of this effect is shown in Fig. 7. For patch scales of 1 to 10 km, the elongation of the patches may be influenced by horizontal velocity shears. The elongation and the distortion of the patches can be, in a number of cases, related to drift and Ekman spiral currents. One such example is illustrated in Fig. 8. These photographs of the patch were obtained in conditions of slight wind and calm sea, when the Ekman depth appeared not to exceed 10-20 m and the dye was penetrating to a depth of 15 m. For stronger winds and rougher sea, i.e. for larger Ekman depths, similar pictures

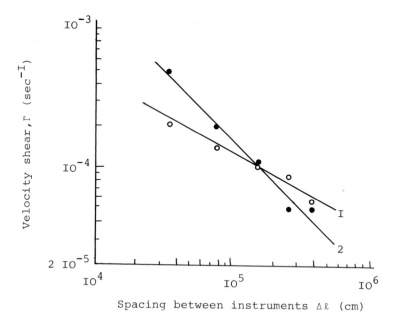

Fig. 5. Variation of the horizontal velocity shear, $\Gamma = \dfrac{\Delta U}{\Delta \ell}$, with the instrument spacing, $\Delta \ell$. Experiment n°1 took place in the Crimea coastal region, experiment n°2 in the Caucasus coastal region.

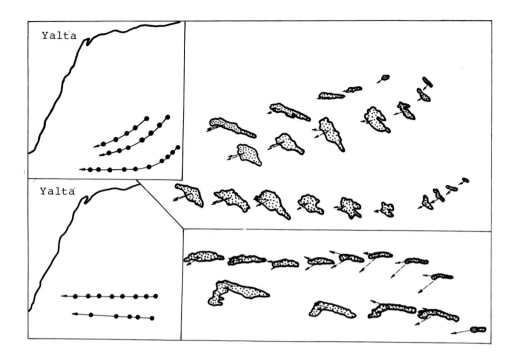

Fig. 6. Schematic representation of the drift of diffusing patches
of dye in the Crimea coastal region. The arrows indicate the wind
and current directions.

of patch distortion were not observed.

Air photographs often show a "streak" structure of the dye field.
These streaks can be related to Langmuir circulations. According to
our estimates the distance between these circulations varied from
5-10 to 20-30 m. This distance is smaller than the one observed in
the open ocean by Assaf et al. (1971). The difference seems to be
related to the fact that a sharp and fairly shallow (20-30 m)
pycnocline was present in the region of our observations and that
the wind velocity was not very large.

Air photography of dye patches can also reveal interesting
phenomena in frontal zones. Such a frontal zone was observed in a
coastal region between transparent waters near shore and a "tongue"
of more turbid waters seaward. Patches of dye were first released
at a distance of 500 m on both sides of the front. The patches
floated rapidly towards the front, they became parallel to the
front in its vicinity, and they disappeared rapidly at the frontal
line. The velocity at which the patches approached the front reached 5-10 cm/sec. Based on the rates of disappearance of the

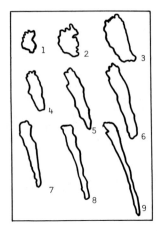

Fig. 7. Elongation of a dye patch caused by vertical velocity shear.

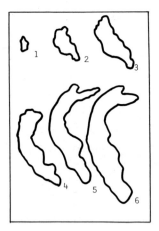

Fig. 8. Elongation and distorsion of a dye patch under the action of drift currents (the "head" of the patch is its southern tip).

patches, one can conclude that the vertical velocities in the
frontal zone had the same order of magnitude. In a second experiment, we released 2 patches seaward of the front at distances of
1.3 and 3 miles. At these distances, the velocities at which the
patches approached the front were small, and during 4-5 hours of
observation the patches were not absorbed by the front. Obviously,
aerial photography of dye patches in frontal zones can provide
valuable data on the structure and the kinematics of these interesting and important features. The wide application of remote methods
using dye as a tracer can provide detailed information on the
interactions of currents, wave motions and turbulence in various
hydrometeorological and orographical situations. The application
of only contact methods of measurements would be insufficient for
such an investigation in a number of cases.

REFERENCES

Assaf,G.,Gerard,R. and Gordon,H., 1971. Some mechanisms of oceanic
    mixing revealed in aerial photographs. J.Geophys.Research, 76,
    6550-6572.
Djuric, D. and Leribaux, H.R., 1974. On the determination of tur-
    bulent diffusivity in shallow waters by serial photography of
    floating markers. Limnology and Oceanography, 19, 138-144.
Fucuda, V., Ito,N. and Sakahishi, S., 1964. Dispersion phenomena
    in coastal areas. Proc. of the 2-nd Inter. Conf. held in Tokyo,
    3, E.A.Pearson (Editor). Pergamon Press.
Ichiye, T., 1965. Diffusion experiments in coastal waters using
    dye techniques. Proc.Symp. Diffusion in Oceans and Fresh Waters
    at Lamont Geol. Obs., Palisades, N.Y. 54-67.
Ichiye, T. and Plutchak, N.E., 1966. Photodensitometric measure-
    ments of dye concentration in the ocean. Limnology and Oceano-
    graphy, II, 364-370.
Ichiye, T., 1967. Upper ocean boundary layer flow determined by
    dye diffusion. Physics of Fluids Supplement, 270-277.
Ito, N., 1964. On the small-scale horizontal diffusion near the
    coast. J. Ocean. Soc. of Japan, 19, 182-186.
Ito, N. and Fucuda, M., Tanicawa, V., 1966. Small-scale horizontal
    diffusion near the coast. Disposal Radioactiv. Wastes in the
    Seas, Oceans and Surface Waters, Vienna.
James, W. and Burgess, F.J., 1970. Ocean outfall dispersion.
    Photogramm. Engineering, 36, 1241-1250.
Nanniti, T., 1964. Some observed results of oceanic turbulence.
    Science on Oceanography, 211-215.
Pritchard, D.W., Okubo, A. and Carter, H., 1966. Observation and
    theory of eddy movement and discussion of an introduced tracer
    material in the surface layers of the sea. Disposal Radioact.
    Wastes in the Seas, Oceans and Surface Waters, Vienna, 397-424.
Schott, F., Ehlers, M., Hubrich, L. and Quadfasel, D., 1978. Small
    scale diffusion experiments in the Baltic surface-mixed layer
    under different weather conditions. Dt. hydrogr. Z., 31, 195-
    215.

SURFACE-WAVE EXPRESSION OF BATHYMETRY OVER A SAND RIDGE

C. GORDON, D. GREENWALT and J. WITTING
Naval Research Laboratory, Washington, DC 20375

ABSTRACT

Satellite SAR and airborne SLAR produce strong surface imagery of bottom features in shallow seas. Lagrangian and Eulerian current measurements have been made in the vicinity of a sand ridge that generates a visual and remotely-sensible wave manifestation known as a rip. The phenomenon is interpreted in terms of wave-current interaction in which the propagation of shorter waves is blocked by an adverse current.

INTRODUCTION

During the past decade wave patterns on the ocean surface have been observed using coherent imaging radar (Brown et al., 1973, 1976; Moskowitz, 1973; Larson and Wright, 1974). Progress in this field has been recently reviewed in detail by Alpers, et al. (1981) This radar imagery is of great interest to oceanographers because the changes in surface roughness responsible for the effect are related to physical oceanographic and meteorological parameters. For example, variations in coherent radar backscatter have been attributed to local wind, surface fronts, ocean swell, internal waves, surface slicks, currents, island shadowing and subsurface topography. The influence of the last of these sources affecting radar imagery is the subject of the work described here. This report is an initial interpretation of the surface-wave manifestation of a subsurface topographic feature (Phelps Bank, Nantucket Shoals) based on photographic, bathymetric and current measurements obtained during a field exercise aboard the USNS HAYES, July, 1982.

BACKGROUND

The first investigators to note the appearance of subsurface
topographic features in the sea-surface imagery of coherent radar
(Side-looking Airborne Radar, SLAR) were DeLoor and Brunsveld van
Hulten (1978). They reported that the surface wave pattern had the
same crest direction and wavelenght as the sand waves on the
bottom of the North Sea where their measurements were taken. They
tentatively and qualitatively attributed the effect to "interfe-
rence of the tidal current with the bottom topography, producing
a very weak wave-like pattern which modulated the capillary waves
in combination with velocity fluctuations at the sea surface".
With regard to the present work they made one particularly relevant
observation, that is, no surface manifestations of subsurface
sand-wave patterns were detected with turning of the tide, but
rather only when diurnal current was present. The implication is
that the topographic effect on the surface-wave pattern is probably
current related. The influence of near-surface currents on the
wavelenghts, amplitudes, speeds and directions of surface waves
has been observed and treated theoretically for a long time (Unna,
1942; Johnson, 1947; Taylor, 1955; Ursell, 1960; Longuet-Higgins
and Stewart, 1960, 1961). The appearance of current effects in the
surface wave field as imaged by coherent radar was first noted by
Brown et al. (1976) but no quantitative explanation was offered.
Tidal currents in shallow seas and the influence of bottom topography on their radar images has also been discussed in a qualitative way by DeLoor (1981). The explanation proposed by DeLoor
(1981) is that "through the tidal current the dune pattern at the
bottom modulates the capillary and short gravity waves at the
surface which are in resonance with the radar wave". The specula-
tion was based on a comparison of radar imagery with bathymetric
charts of the same area, rather than a detailed analysis. The sea
depth at the location was about 25 m and the "wave height" of the
subsurface sand dunes was about 4 m. The tidal currents involved
ranged between 0.5 and 1 msec^{-1}.

SLAR imagery of the sea surface in the Southern Bight of the
North Sea off the coast of Holland has been reported by McLeish et
al. (1981) to reflect sea-floor bed forms such as sand ridges and
sand waves. The sand ridges were the order of 500 m wide and 10 to
25 m below the surface at their crests (Buitenbank, Schouwenbank
and Bollen van Goeree). They occur in an area where currents are

dominated by rotary tidal components in the range of 0.25 - 0.50
msec^{-1}. McLeish et al. (1981) conclude that water depth is the
controlling variable responsible for changes in the coherent radar
imagery. Basing their interpretation on the "hydrodynamic effect"
of ocean waves on radar as proposed by Valenzuela (1978), McLeish
et al. (1981) assume that the Bragg waves (order of cm) on the
surface that are "seen" by the radar are changed in shape by convergence and divergence of the current flow as it passes over the
topographic feature. They suggest that the surface convergence
where the sea is deeper tends to increase the density of wave
energy whereas divergence over the top of the shallower features
will have the opposite effect on waver energy density. It was concluded that the surface manifestation requires two conditions :
1) water flow across a sufficient relative depth change and 2) wind
speed sufficient to generate Bragg waves. These investigators also
observed that the appearance and disappearance of surface patterns
from bottom features depended on the direction of the tidal
currents. The features that disappeared in the radar imagery were
those roughly parallel to the current direction. As in the earlier
work, the bathymetry effects on surface waves are associated with
current flow. Some of the most spectacular indications of surface
wave patterns reflecting bottom topography have been observed
with the SAR (Synthetic Aperture Radar) aboard the SEASAT satellite, particularly in the Nantucket shoals area. As yet there are no
detailed analyses or interpretations of this data. As pointed out
by Phillips (1981) in his discussion of the SAR imagery and the
interaction of short gravity waves with surface currents, there has
been little systematic study of the spatial variability of surface
currents induced by steady or tidal flow across irregular shallow-
bottom topography.

Figure 1 is an illustration of SAR imagery in the Nantucket-Cape
Cod area. The location of the subsurface sand ridge called Phelps
Bank that is the object of this study is shown in the figure. As
Kasischke et al. (1980) have pointed out, approximately 80% of the
surface features in this imagery were over or near distinct bottom
features. The abrupt change in the character of the surface wave
field in the vicinity of Phelps Bank is clearly seen from the air.
Figure 2 is an aerial photograph that shows this sharp demarkation.

The fact that surface currents interact with gravity waves to
change their properties has been the subject of numerous studies.

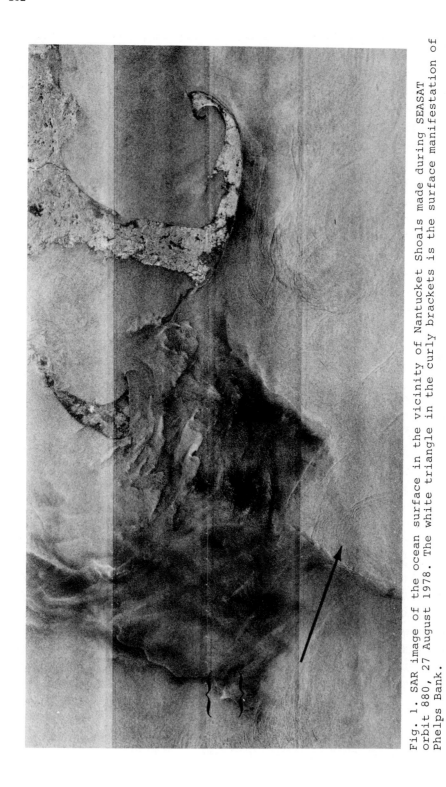

Fig. 1. SAR image of the ocean surface in the vicinity of Nantucket Shoals made during SEASAT orbit 880, 27 August 1978. The white triangle in the curly brackets is the surface manifestation of Phelps Bank.

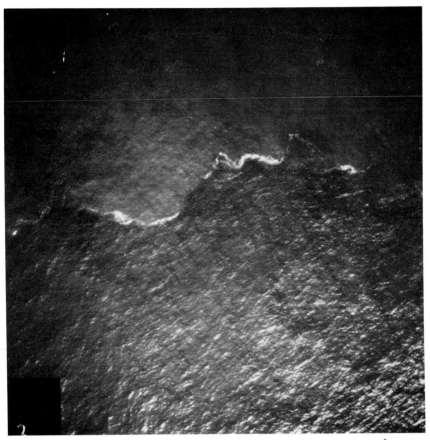

Fig. 2. An aerial photograph taken near Phelps Bank from an altitude of about 460 m. The camera focal length was 15 cm. A sharp change in surface waves is clearly evident.

Unna (1942) examined the case in which on-shore waves encounter ebbing tidal currents from an estuary at approximately 180° between the directions of current flow and the wave propagation. He found that when waves traveling at speed V_o meet a current moving at speed C the waves will break when $C = -1/4\ V_o$ regardless of the initial steepness of the wave. Thus in a tidal race or on a bar there can be standing lines of breaking waves and to the leeward of the position where $C = -1/4\ V_o$ the sea will be relatively calm. For example, he found that waves with slack-water wavelengths of 30 m are essentially blocked by a counter current of 1.8 msec^{-1}. Johnson (1947) later demonstrated that wave blocking, breaking and refraction may occur even when the waves do not advance directly into the current. His analysis implied that when waves encounter a current of 1.0 msec^{-1}, all the waves with speeds less than 10 msec^{-1} (7 sec period) that enter the current at an angle greater than 58° will break and will therefore be unable to cross the current. Johnson (1947) demonstrated the wave-current interaction with a remote sensing experiment using aerial photographs of waves passing through ebb and flood tidal currents at the entrance of Humboldt Bay, California. Taylor (1955) examined wave-current interaction theoretically for the practical purpose of utilizing the wave blocking phenomenon in designing a "current-flow" breakwater for harbor protection. He found that the effect was not strongly depen- dent on the vertical distribution of the current flow. That is, the wave stopping efficiency of a uniform stream is equivalent to that of a stream with a uniform velocity gradient provided the total flow of inertia is the same. An interesting example of Taylor's (1955) results is that if a 4.6 msec^{-1} current is present over a sufficiently long distance, it would only have to be 0.5 m deep to stop waves with 30 m wavelengths. More recent work by Longuet- Higgins and Stewart (1960, 1961) and Ursell (1960) has treated the problem of changes in the form of short gravity waves by surface\ currents in rather rigorous mathematical detail.

FIELD MEASUREMENTS

On July 14, 1982, 1400 U.T. the USNS HAYES was conducting measurements of wave spectra in the vicinity of Phelps Bank, Nantucket Shoals (40°50'N-60°20'W). Two wave-recording buoys were deployed, one free and one tethered to the ship by about 100 m of cable. The ship was adrift with the exception of occasional slow maneuvers to maintain slack cable to the tethered buoy. The ship was carried in a south-westerly direction by the tidal current at a relatively rapid speed (0.8-1.5 msec^{-1}) into an area labeled on the charts as Asia Rip (NOAA, 1979). It should be noted that the designation "rip" derives from ripple and implies an area of water made rough by opposing tides or currents. Its presence on a navi- gational chart also indicates that it is a visible and relatively permanent surface feature. The East to West course of the USNS HAYES over Phelps Bank during the drift into Asia Rip is shown as Track A in Figure 3. As anticipated from the navigational charts, the rip was indeed clearly visible with a rather sharp line of demarcation separating an area of considerable wave activity from the relatively calm ambient sea. The photograph in Figure 4 provides a qualitative indication of the change in sea state approaching the rip. The change was roughly equivalent to moving from a sea at about Beaufort scale 2 with small wavelets and a glassy appearance to one at Beaufort scale 3 with large wavelets and some breaking crests. The wind at a height of 10 meters was approximately 5 msec^{-1} at the time. Since neither the meteorological or hydrographic conditions changed significantly as the ship passed into Asia Rip, it was tentatively assumed that the sharp contrast in surface waves was a dynamic effect, probably related to wave interaction with current flow. Because the location of the rip was near the southern end of a prominant, 9 km-long bedform (Phelps Bank, 40°50'N-69°20'W) and inasmuch as the ship course (Track A) showed no sharp directional changes indicating horizontal current shearing it was decided to pursue the hypothesis that the variations in the near-surface current responsible for the change in wave pattern are related to the bottom topography.

During the at-sea exercise aboard the USNS HAYES, a chart recorder maintained a continuous graphical plot of the depth as measured by a sonic fathometer. Time marks were placed on the bathymetry record by watch standers at intervals varying from a few minutes to a few hours, which makes it possible to calculate the depth at a given time based on the chart recorder speed. Furthemore, navigational information from the LORAN-C system was averaged over each minute and logged by computer, providing a time series record of both ship speed and position. By correlating the records for the appropriate times during the cruise it was possible to derive current and bathymetry information at Phelps Bank in general and to relate it to the crossing of Asia Rip in particular.

The navigational and bathymetric data pertaining to the "drift" into Asia Rip are given in Table 1 and shown graphically in Figure 5. The bank profile in the figure represents the recorded depths along Track A plotted in terms of west longitude only. The approximate position of the USNS HAYES at the time of the photograph in Figure 4 is illustrated with some artistic license. The location of the change in surface wave pattern relative to the subsurface topography is presented schematically. The least interpretable information in the figure is the current or ship drift. As mentioned earlier, the ship at the time, was maintaining position relative to a tethered buoy. It was primarily in a drifting regime, but there is a possibility that some occasional maneuvering took place to accommodate the deployed buoy. Since this particular analysis was not anticipated at the time, no record was kept of ship propeller rotations so there is some ambiguity in the inter- pretation of the ship speed data. The LORAN-C positions in the area are quite accurate (less than $\pm$ 100 m) but the question of the contribution of short-term ship maneuvers to the speed curve remains open. In a qualitative sense, the significant correlation to be gained from Figure 5 is that the change in surface wave pattern coincides spatially with the lee edge of the bank and what appears to be a drop of 30-40 cmsec^{-1} in current speed (assuming that the ship was moving as a Lagrangian drifter at the time).

Later in the cruise (July 21, 1982) there was an opportunity to repeat an East-to-West ship drift across the bank (Figure 3, Track B) and eliminate the ambiguity associated with possible ship maneuvers. This was done by stopping all engines. Table 2 lists navigational and bathymetric data at two minute intervals for this pass across Phelps Bank. The basic premise in this measurement is that the USNS HAYES (keel depth 5.8 m) acts as a Lagrangian drifter when not under power, that is, the progressive series of LORAN-C fixes recorded aboard ship represent motion of the ocean current only. For this premise to be valid, it must be assumed that no other forces are acting on the ship. The only other force of any consequence would be the wind acting on the super-structure of the ship as a sail. Table 3 provides information on the relationship of the ship course, ship heading and the wind while traversing Track B. It is seen from the table that both the ship heading and the wind velocity remained relatively constant during the drift across the bank, that is, the wind remained on the starboard quarter. This is interpreted to mean that the wind effect is cons- tant and if it has any influence, it will be a steady southerly displacement of the ship track. Such an effect could be subtracted out if the ship movement were known as a function of wind speed alone. Therefore, based on this rather qualitative argument, it is considered that the ship drift is probably linearly related to the current.

The information in Tables 2 and 3 is summarized and correlated graphically in Figures 6 and 7. The bathymetric profile of the bank along the drift track is shown in both figures for common reference. Figure 6 shows a segment of Track B, plotting LORAN-C

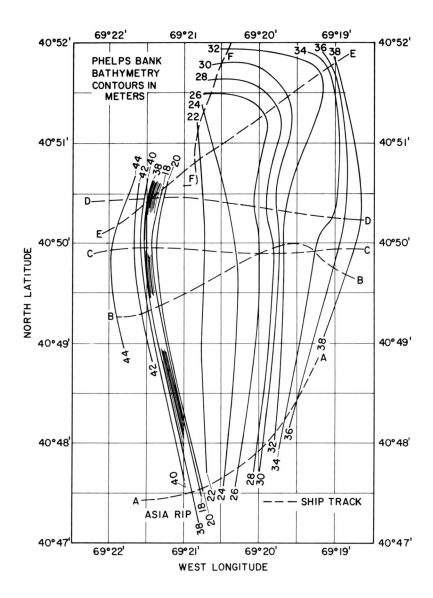

Fig. 3. The general topography of Phelps Bank based on bathymetric measurements obtained during six ship crossings (Track A-F). Asia Rip is located near the southern end of the bank.

positions at 4 or 8-minute intervals. The "wind rose" for the duration of the drift is also shown in the figure. It should be noted that the wind remains constant within about ± 25° in direction ± 1.5 msec^{-1} in speed. Figure 7 is equivalent to Figure 5, however, there were no accompanying surface wave observations

Table 1. Depths, positions and ship speeds along Track A, July 14, 1982

Time (U.T.)	Speed (MSEC^{-1})	Depth (M)	N. Latitude	W. Longitude
1420	1.62	36.0	40° 48.45'	69° 19.49'
1422	1.73	35.5	40° 48.34'	69° 19.57'
1424	1.74	34.0	40° 48.25'	69° 19.65'
1426	1.65	33.0	40° 48.17'	69° 19.73'
1428	1.50	31.0	40° 48.11'	69° 19.82'
1430	1.42	30.5	40° 48.04'	69° 19.90'
1432	1.36	29.5	40° 47.97'	69° 19.97'
1434	1.25	28.5	40° 47.91'	69° 20.04'
1436	1.22	27.5	40° 47.85'	69° 20.12'
1438	1.29	26.5	40° 47.80'	69° 20.21'
1440	1.34	25.5	40° 47.75'	69° 20.31'
1442	1.34	24.5	40° 47.70'	69° 20.40'
1444	1.28	24.0	40° 47.65'	69° 20.49'
1446	1.23	23.5	40° 47.61'	69° 20.58'
1448	1.19	23.5	40° 47.57'	69° 20.67'
1450	1.20	22.5	40° 47.52'	69° 20.76'
1451.5	----	21.0	Shallowest	Point
1452	1.09	24.0	40° 47.50'	69° 20.82'
1454	0.84	38.0	40° 47.48'	69° 20.89'
1456	0.89	37.5	40° 47.49'	69° 20.99'
1458	1.13	37.5	40° 47.49'	69° 21.09'
1500	1.12	37.0	40° 47.48'	69° 21.18'
1502	1.00	37.0	40° 47.47'	69° 21.26'
1504	0.89	37.0	40° 47.46'	69° 21.33'
1506	0.81	----	40° 47.45'	69° 21.39'
1508	0.28	----	40° 47.44'	69° 21.46'
1510	0.25	----	40° 47.44'	69° 21.53'
1512	0.74	----	40° 47.44'	69° 21.59'
1514	0.76	----	40° 47.43'	69° 21.66'

because it was dark and foggy during the pass (Local EDT is 4 hrs earlier than U.T.). The observation of primary interest is the current variation relative to the topography of Phelps Bank. In common with Figure 5, there is a marked change in current (ship drift) speed correlated with the leeward edge of the bank. However, there are fewer large fluctuations in current speed than was the case during the "drift" into Asia Rip (Figure 5). This may be attributable to operating with engines stopped or may indicate that there are fewer turbulent eddies with dimensions equivalent to several shiplengths in this area. A particularly noticeable dynamic feature is the rather abrupt change in the East-West current speed (U) gradient around Longitude 69°21.15'. dU/dX undergoes a change of approximately a factor of five at this point. Upstream of the break point dU/dX is 0.14×10^{-3} sec^{-1} while the downstream value is 0.66×10^{-3} sec^{-1}. There is some indication that the current changes are attributable to the bank as such. For example, Figure 7 also includes a current that would be predicted in the absence of any topographic feature. This current prediction

Fig. 4. A photograph of the sea surface in the vicinity of Asia Rip showing the sharp contrast in surface wave activity. The field of view is directed from Phelps Bank towards the lee side of the feature (west).

is based on Lagrangian measurements that were made west of Phelps Bank during an independent experiment using drogues to follow the trajectory of tidal flow in the vicinity. Even though the predicted current may be less than precise, the implication is that current across the shallow part of the bank is faster than would be expected in the absence of the topography while the current in the lee of the sand ridge is slower than anticipated.

In the course of the remote sensing field exercise there were no other occasions when Phelps Bank was crossed in a drifting mode. During the other East-West passes over the bank shown in Figure 3 (Track C,D,E), the ship was under power. If it is assumed that the USNS HAYES was moving through the sea at constant speed (RPM), two of these passes (C and D) do provide some supplementary information on current variation over the bank. During the course of Track C

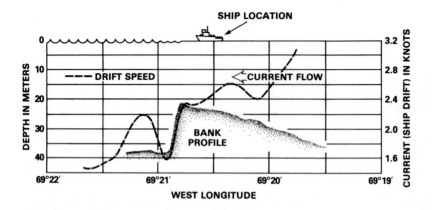

Fig. 5. A schematic diagram showing the spatial relationship of Phelps Bank, the surface-wave changes and the USNS HAYES at the time the photograph in Fig. 4 was taken. The ship drift speed across the bank is also illustrated. It should be kept in mind that some of the variation in ship speed may be a result of deliberate maneuvers.

(ship traveling West to East), the predicted current was flowing East to West (280°) at about 0.5 msec^{-1}, that is, the current opposed the ship motion. Data on the ship speed and bathymetry for this track are given in Table 4. The observation of interest is that there is an approximately 0.23 msec^{-1} increase in ship speed as it passes over the lee edge of the bank, an indication of a slower surface current at that location. This is in qualitative agreement with the results of the measurements when the USNS HAYES was in a drifting mode. Analogous information for Track D (Fig. 3) is given in Table 5. Here the ship is crossing the bank under power from East to West. The predicted tidal current at the time flows from West to East (125°) and as in the case of Track C, opposes the ship motion. As seen from Table 5, the ship appears to slow down by approximately 0.34 msec^{-1} as it passes over the "shallows" of the bank. Since the current opposes the ship's progress, this indicates that the faster current is over the bank with slower currents at both the upstream and downstream edges of the bank. Although the interpretation of ship speeds under power as current indicators are less reliable than observations in the drifting mode, it should be noted that they are consistent with the drift data.

The influence of the sand ridge on the local flow pattern is also confirmed by a comparison of simultaneous Eulerian and Lagrangian current measurements made near the site. Figure 8 is a

Table 2. Depths, positions and ship speeds along Track B,
July 21, 1982

Time (U.T.)	Speed (MSEC^{-1})	Depth (M)	N. Latitude	W. Longitude
0943	1.09	29.0	40° 49.96'	69° 19.74'
0945	1.03	28.0	40° 49.93'	69° 19.82'
0947	1.01	25.5	40° 49.91'	69° 19.90'
0949	.93	26.0	40° 49.89'	69° 19.97'
0951	.85	25.0	40° 49.88'	69° 20.04'
0953	.82	24.0	40° 49.87'	69° 20.13'
0955	.91	24.0	40° 49.84'	69° 20.21'
0957	.92	23.5	40° 49.81'	69° 20.26'
0959	.88	23.5	40° 49.77'	69° 20.32'
1001	.87	23.0	40° 49.74'	60° 20.39'
1003	.82	23.0	40° 49.71'	60° 20.45'
1005	.86	23.0	40° 49.69'	60° 20.52'
1007	.92	23.5	40° 49.66'	60° 20.60'
1009	.83	23.5	40° 49.63'	60° 20.65'
1011	.77	23.5	40° 49.61'	69° 20.71'
1013	.65	24.0	40° 49.59'	69° 20.77'
1015	.67	23.5	40° 49.58'	69° 20.82'
1017	.76	24.5	40° 49.55'	69° 20.87'
1019	.71	23.5	40° 49.52'	69° 20.94'
1021	.65	22.5	40° 49.50'	69° 20.98'
1023	.65	22.0	40° 49.49'	69° 21.03'
1025	.67	23.5	40° 49.48'	69° 21.10'
1027	.70	24.5	40° 49.45'	69° 21.16'
1029	.69	23.0	40° 49.42'	69° 21.20'
1031	.62	23.5	40° 49.41'	69° 21.24'
1033	.50	22.5	40° 49.41'	69° 21.28'
1035	.50	18.4	40° 49.40'	69° 21.33'
1037	.55	31.5	40° 49.37'	69° 21.37'
1039	.50	37.0	40° 49.36'	69° 21.41'
1041	.36	38.5	40° 49.35'	69° 21.42'
1043	.32	39.5	40° 49.34'	69° 21.46'
1045	.35	40.5	40° 49.33'	69° 21.49'
1047	.35	41.5	40° 49.32'	69° 21.52'
1049	.22	41.0	40° 49.32'	69° 21.54'
1051	.21	41.5	40° 49.32'	69° 21.55'
1053	.26	41.5	40° 49.31'	69° 21.58'
1055	.26	42.0	40° 49.29'	69° 21.59'
1057	.21	41.5	40° 49.30'	69° 21.61'
1059	.10	42.0	40° 49.30'	69° 21.63'
1101	.26	42.0	40° 49.29'	69° 21.66'
1103	.26	42.0	40° 49.29'	69° 21.69'
1105	.26	42.0	40° 49.27'	69° 21.70'
1107	.15	42.0	40° 49.28'	69° 21.71'
1109	.10	42.5	40° 49.28	69° 21.73'
1111	.15	42.0	40° 49.28'	69° 21.75'
1113	.21	42.0	40° 49.27'	69° 21.77'
1115	.21	42.0	40° 49.27'	69° 21.78'
1117	.05	41.0	40° 49.28'	69° 21.78'
1119	.05	40.5	40° 49.28'	69° 21.79'
1121	.00	40.5	40° 49.28'	69° 21.81'
1123	.10	40.0	40° 49.29'	69° 21.84'
1125	.10	40.0	40° 49.29'	69° 21.85'
1127	.05	40.0	40° 49.30'	69° 21.84'
1129	.15	40.0	40° 49.31'	69° 21.84'
1129	.15	40.0	40° 49.31'	69° 21.84'

Table 3. Wind conditions and ship heading during a pass over
Phelps Bank (July 21, 1982)

Time (U.T.)	Course (Deg)	Ship Hdng	Wind Speed	Wind Dir
0945			7.3	2.8
0947	246			
0949	250			
0951	256			
0953	257			
0955	254			
0957	250			
0959	237			
1001	233		7.8	326
1003	236			
1005	245			
1007	244			
1009	241			
1011	240			
1013	236			
1015	244	190°	9.5	347.4
1017	245	220°		
1019	237	235°		
1021	238	230°		
1023	248	210°		
1025	257	200°		
1027	242	220°		
1029	236	240°		
1031	243	240°	9.9	354.5
1033	249	220°		
1035	257	220°		
1037	248	235°		
1039	237	235°		
1041	248			
1043	258	220°		
1045	236	240°	9.8	3.9
1047	243	250°		
1049	251	230°		
1051	267	220°		
1053	256	240°		
1055	242	260°		
1057	231	250°		
1059	251	240°		
1101	272		10.4	7.9
1103	251	270°		
1105	234	270°		
1107	246	250°		
1109	286	240°		
1111	264	250°		
1113	249			
1115	257	270°	10.3	13.5
1117	281	250°		
1119	335	240°		
1121	265	250°		
1123	266	260°		
1125	267	270°		
1127	312	260°		
1129	248	240°		

Note: 1) Ship engines stopped at 0938
2) Wind speed is in meters per second
3) Wind direction and ships course and heading are in degrees
4) Wind speeds and directions are 15 minute averages

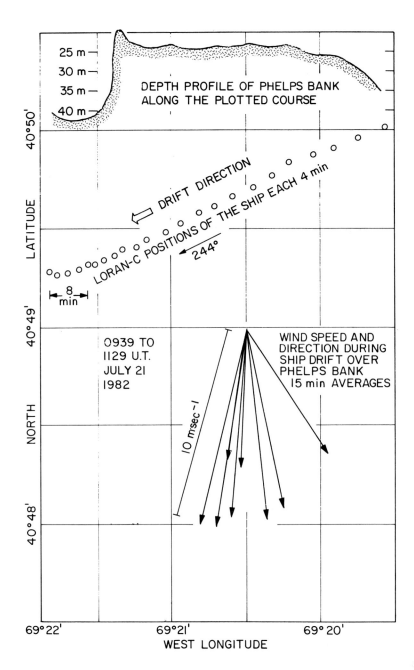

25 m —
30 m —
35 m —
40 m —

DEPTH PROFILE OF PHELPS BANK
ALONG THE PLOTTED COURSE

DRIFT DIRECTION

LORAN-C POSITIONS OF THE SHIP EACH 4 min

244°

8 min

0939 TO
1129 U.T.
JULY 21
1982

WIND SPEED AND
DIRECTION DURING
SHIP DRIFT OVER
PHELPS BANK
15 min AVERAGES

10 msec⁻¹

LATITUDE

NORTH

40°50'
40°49'
40°48'

69°22'   69°21'   69°20'
WEST LONGITUDE

Fig. 6. Ship drift across Phelps Bank (Track B, Fig. 3). Winds during the
drift and the bathymetric profile along the drift path are also
shown.

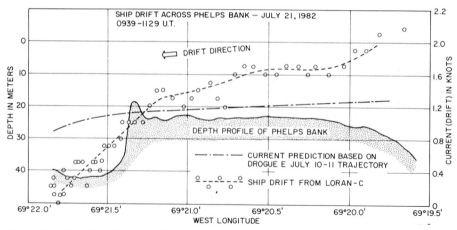

Fig. 7. The drift speed of the USNS HAYES along the path shown in Fig. 6 (Track B. Fig. 3).

computer plot showing the trajectory of a current-following drogue about 8 km west of Phelps Bank compared to the progressive vector diagram derived from a recording current meter moored at the sand ridge (40°50.07'N - 69°19.81'W). The drogue and the current meter were at approximately the same depth (5-6 m). As seen from the figure, there is an almost two-fold increase in the East-West component of the tidal current as it passes over the sand ridge perpendicular to its major axis. We interpret this enhancement of current flow over Phelps Bank as a topographic effect, that is, an acceleration due to the vertical reduction in "channel depth".

Based on these field measurements which are admittedly somewhat limited in both quantity and degree of reliability, the following conclusions can be drawn. Relatively large changes in surface current speed (25 to 50 cm/sec) and current gradient (0.20 - 1.5 x 10^{-3} sec^{-1}) occur in the vicinity of this topographic feature, with faster currents over the shallow portion of the bedform. The most significant variations in current are spatially correlated with the limits of the bank. The change at the Western edge of Phelps Bank was particularly noticeable and on at least one occasion was clearly associated with a sharp contrast in the appearance of the surface wave pattern. It seems reasonable to attempt to relate the surface wave variations to the bathymetry through the mechanism of wave-current interaction. This will be explored in the discussion to follow.

Table 4. Depths, positions and ship speeds along Track C,
July 14, 1982

Time (U.T.)	Speed (MSEC^{-1})	Depth (M)	N. Latitude	W. Longitude
1715	4.97	44.5	40° 49.89'	69° 22.48'
1716	4.72	44.0	40° 49.91'	69° 22.31'
1717	4.54	43.0	40° 49.93'	69° 22.15'
1718	4.53	43.0	40° 49.94'	69° 21.94'
1719	4.65	43.0	40° 49.96'	69° 21.73'
1720	4.72	39.0	40° 49.97'	69° 21.53'
1720.5	----	19.5		
1721	4.76	23.5	40° 49.97'	69° 21.33'
1722	4.65	24.0	40° 49.95'	69° 21.15'
1723	4.58	23.0	40° 49.95'	60° 20.95'
1724	4.51	24.0	40° 49.94'	60° 20.76'
1725	4.48	22.0	40° 49.93'	60° 20.57'
1726	4.52	26.5	40° 49.91'	60° 20.37'
1727	4.54	24.0	40° 49.91'	60° 20.18'
1728	4.54	26.5	40° 49.90'	69° 19.98'
1729	4.54	28.0	40° 49.91'	69° 19.78'
1730	4.52	30.5	40° 49.91'	60° 19.57'
1731	4.52	31.0	40° 49.92'	60° 19.39'
1732	4.55	35.5	40° 49.93'	69° 19.19'
1733	4.54	35.0	40° 49.94'	69° 19.00'
1734	4.57	41.0	40° 49.94'	69° 18.81'
1735	4.59	42.5	40° 49.95'	69° 18.60'

Table 5. Depths, positions and ship speeds along Track D,
July 19, 1982

Time (U.T.)	Speed (MSEC^{-1})	Depth (M)	N. Latitude	W. Longitude
1530	4.22	48.0	40° 50.18'	69° 18.10'
1531	4.29	39.0	40° 50.20'	69° 18.28'
1523	4.33	39.0	40° 50.22'	69° 18.47'
1533	4.36	38.0	40° 50.23'	60° 18.65'
1534	4.32	37.0	40° 50.25'	60° 18.82'
1535	4.18	34.0	40° 50.25'	60° 19.00'
1536	4.10	33.0	40° 50.27'	60° 19.17'
1537	4.10	37.0	40° 50.29'	60° 19.35'
1538	4.11	34.0	40° 50.31'	69° 19.53'
1539	4.08	32.0	40° 50.32'	69° 19.69'
1540	4.08	29.0	40° 50.35'	60° 19.87'
1541	4.07	27.0	40° 50.36'	69° 20.04'
1542	4.03	25.0	40° 50.38'	69° 20.20'
1543	3.98	25.0	40° 50.39'	60° 20.37'
1544	3.98	26.5	40° 50.41'	69° 20.54'
1545	3.93	22.0	40° 50.43'	69° 20.72'
1545.3	----	19.5		
1546	3.98	25.0	40° 50.44'	69° 20.89'
1547	4.06	21.0	40° 50.47'	69° 21.07'
1547.3	----	18.0		
1548	4.12	21.0	40° 50.46'	69° 21.24'
1548.2	----	18.0		
1549	4.15	37.5	40° 50.45'	69° 21.43'
1550	4.20	43.0	40° 50.44'	69° 21.62'
1552	4.32	45.0	40° 50.43'	60° 22.00'
1553	4.39	45.0	40° 50.42'	69° 22.19'
1554	4.38	43.0	40° 50.41'	69° 22.38'
1555	4.38	44.0	40° 50.40'	69° 22.56'
1556	4.36	42.0	40° 50.39'	60° 22.75'

DISCUSSION

For purposes of discussion, we will consider the measurements described in terms of simplified fluid mechanics. Figure 9 is a two-dimensional picture of the problem. The figure does some injustice by deletion to the complicated oceanographic situation but includes its essential features. A strong current in relatively shallow water flows over a bathymetric feature and becomes less strong in the deeper water. The water surface in the lee of the feature is covered with waves, some breaking. The obvious waves are shorter than the water depth ($\sim$ 20-40 m), so we take the relevant water waves to be deep water waves.

Let us first consider, in the simplest possible way, how the currents may be modifying a pre-existing wave field to the extent observed, namely an apparent blocking of the shorter waves. The approach is similar to that of Unna (1942) and is purely kinematic. The velocity C_1 of deep-water waves moving over still water is proportional to the square root of their wavelenght.

$$C_1 = \lambda_1^{1/2} \left[\frac{g}{2\pi}\right]^{1/2} \tag{1}$$

If the waves encounter an opposing current, u, the wave velocity relative to the moving current C_r will be

$$C_r = \lambda_2^{1/2} \left[\frac{g}{2\pi}\right]^{1/2} \tag{2}$$

where λ_2 is the wave length associated with velocity C_r. The new velocity over the bottom is $C_2 = C_r - u$. From the velocity equations, it can be seen that the ratio

$$\frac{\lambda_2}{\lambda_1} = \left(\frac{C_r}{C_1}\right)^2 \tag{3}$$

However, the period of the waves relative to the bottom does not change as they propagate into the current, so

$$T = \frac{\lambda_1}{C_1} = \frac{\lambda_2}{C_2} = \frac{\lambda_2}{C_r - u} \tag{4}$$

Two aspects are worth noting, that is

$$\lambda_2 = \lambda_1 \frac{C_r - u}{C_1} \tag{5}$$

or the new wavelength is shorter than that in deep water and that the ratio

$$\frac{\lambda_2}{\lambda_1} = \frac{C_r}{C_1} - \frac{u}{C_1} = \left(\frac{C_r}{C_1}\right)^2 \tag{6}$$

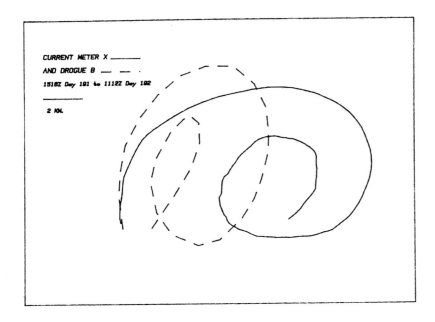

Fig. 8. Comparison of a Lagrangian drogue trajectory 8 km west of Phelps Bank with a progressive vector diagram from a moored current meter on the bank for the same period of time.

from the Equation (3). This yields a quadratic equation in $\left(\frac{C_r}{C_1}\right)$.

$$\left(\frac{C_r}{C_1}\right)^2 - \frac{C_r}{C_1} + \frac{u}{C_1} = 0 \tag{7}$$

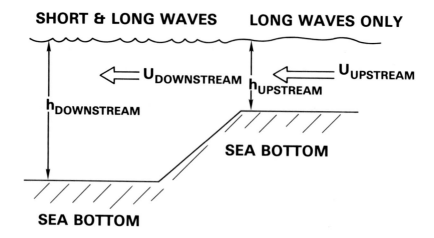

Fig. 9. A schematic diagram of the fluid dynamic conditions at Phelps Bank.

having the solution

$$\frac{C_r}{C_1} = \frac{1}{2} \pm \sqrt{\frac{1}{4} - \frac{u}{C_1}} \tag{8}$$

If $\frac{u}{C_1} = \frac{1}{4}$, $\frac{C_r}{C_1} = \frac{1}{2}$ or $\frac{1}{2} C_r = u$. However, since the group velocity or speed of energy propagation is $\frac{1}{2} C_r$ in the current, we have the situation where the energy propagation velocity is equal and opposite to the current and the waves with still-water speeds of 4 times the current speed or less are blocked.

Unna's (1942) derivation requires : 1) that the currents be steady, 2) that they have gradients only in the direction of the current, 3) that the waves are propagating at a 0° angle of inci-dence into the currents, and 4) that the waves are "deep water waves", i.e., wavelengths shorter than about twice the depth. In the area of our field measurements the currents are certainly not steady over the course of a tidal cycle. Yet over times appropriate for wave blocking, they probably have a component that is steady enough. That the currents have strong gradients in the direction of the current is borne out by the shipboard measurements. The assumption of coalignment of wave and current directions, however, is at best speculative and at worst unwarranted. For this reason, we have derived a somewhat more general kinematic theory that

removes this restriction, while retaining the assumptions of stea-
diness and normal current gradients for simplicity, as well as the
limitation to deep water waves (well borne out by the observations).

We choose the coordinate system in which the currents are most
likely to be steady, that is, geographical coordinates, where the
bathymetric feature causing current gradients is fixed. The dispersion relation for infinitesimal deep water waves riding in a
gradually varying current $u = u(x)$ is :

$$(\omega - k_x u)^2 = gk \tag{9}$$

where ω is the radian frequency of the waves, k_x is the component
of wave number aligned parallel to the current u, k is the magnitude of the wave number ($2\pi/$ wavelength), and g is the acceleration
of gravity. For the wave field not to generate new waves, and for
waves not to break apart across the current gradient we require :

$$\omega = \omega_o = \text{Const}$$

$$k_y = k_{yo} = \text{Const.}$$

In past work it has been customary to define a reference speed
$c_o = g/\omega_o$, which is the speed of deep water waves having frequency
ω_o when $u = 0$. In terms of the reference speed (9) leads to :

$$\frac{c^2}{c_o^2} - \frac{u}{c_o} \cos\theta = \frac{c}{c_o} \tag{10}$$

where θ is the angle between the direction of the current and the
wave. Equation (10) is quadratic in c/c_o, and gives :

$$\frac{c}{c_o} = \frac{1}{2} + \frac{1}{2} \sqrt{1 + \frac{4u \cos\theta}{c_o}} \tag{11}$$

an expression derived by Jonsson et al. (1970). Jonsson et al.
identify the vanishing of $1 + (4u \cos\theta/c_o)$ as marking wave blocking, and claim that

$$u_{blocking} = - \frac{c_o}{4 \cos \theta} \tag{12}$$

At this value of u, (11) gives $c_{block} = \frac{1}{2}c_o$. Thus (12) gives :

$$-u_{blocking} \cos \theta = \frac{1}{2} c_{blocking} \tag{13}$$

We find that waves not propagating directly into the adverse current are blocked <u>before</u> they reach a current magnitude as large as predicted by (12). We proceed as follows. We nondimensionalize (9) and the $\omega = \omega_o = $ const. $k_y = k_{yo} = $ const condition in the units $\omega_o = g = 1$.
Then (9) becomes

$$(1 - k_y u \cot \theta)^2 = k_y \csc \theta, \tag{14}$$

where k_y is constant along a ray. If, instead of θ, k_x is chosen to be an independent variable, we have :

$$(1 - k_x u)^2 = \sqrt{k_x^2 + k_y^2} \tag{15}$$

which is a quartic equation in $k_x = k_x(u)$. Consequently, there are four k_x's for a particular k_y at each u.

We have solved Equation (14) to find the relationships between u and θ that correspond to a wave train propagating with the normal component of the energy flux into the current. These are displayed in Figure 10. We should,imagine that a wave of frequency ω_o is generated at some angle of incidence at a place where the adverse current is u. For simplicity, imagine that u is a monotonic function of x, whith $|u|$ increasing upward in the figure. Then the train propagates up one of the solid lines of the figure until it is blocked. This blocking occurs when, at constant k_y, $\partial u/\partial \theta = 0$.

$$|u| = \frac{1}{2} \cos \theta [1 - \frac{1}{2} \cos^2 \theta] \tag{16}$$

It can be shown that at this point the component of the group velocity parallel to the current equals -u. Thus, we interpret blocking as occuring when the component of group velocity aligned

with the current matches $|u|$. This differs from the interpretation
of Jonsonn et al. that blocking occurs when the component of
current parallel to the wave propagation direction matches the
group speed.

Above the dashed curve in the figure there can be waves that
have phases traveling toward increasing adverse currents; their
direction of energy propagation, however, is with the current, and
so they cannot be reached from below in the figure.

All waves that start under the dashed curve arrive at the blocking current having angles of incidence less than $\theta = \cos^{-1} 2/3$
$= 35.3°$, and all have nondimensional blocking currents between
0.2500 and 0.2722. Their wave numbers range between 4.0000 and
2.2500. All waves that start from the abscissa i.e., are generated
in an oceanic region where there are no currents, arrive at the
blocking current having angles of incidence less than 17.14°, and
all have blocking currents between 0.2500 and 0.2596. Their wave
numbers range between 4.0000 and 3.3861.

The observations described here are best interpreted at a given
location, where the current has a fixed (and approximately known)
value. If wave blocking is occuring at this location, the waves
that are blocked must have : 1) an extremely narrow frequency
range, 2) a narrow spread of directions near normal incidence to
the current, and 3) a rather narrow range of wavelengths. The
theory given here makes these predictions even if the wave field
has components that originated over a wide range of frequencies
and directions. The bounds on these parameters (in dimensional
units) are :

$$T_{min} = \frac{2\pi}{g} \frac{u}{0.2772}$$

$$T_{max} = \frac{2\pi}{g} \frac{u}{0.2550} \tag{17}$$

so for T measured in seconds and u in meters/sec, wave blocking in
this theory will be observed in the narrow range of periods :

$$2.3554 \; u/(m/sec) < T/sec < 2.5646 \; u/(m/sec) \tag{18}$$

Higher frequency waves (smaller periods) will have been blocked
before reaching the observation site, and lower frequency waves
(longer periods) can pass through if initially infinitesimal.

WAVE BLOCKING TRAJECTORIES

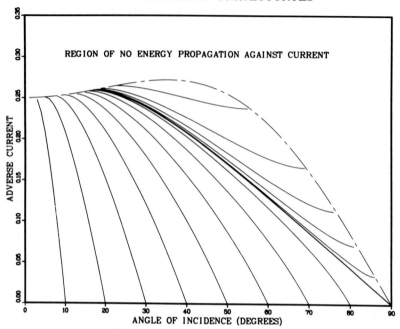

Fig. 10. Wave blocking trajectories. The adverse current is measured in units of g/ω_O, where g is the acceleration of gravity and ω_O is the radian frequency of the wave. A wave moves up a solid curve, turning into the current until blocked by the current. Below the dashed curve waves can propagate against the current; above the dashed curve they cannot.

The ranges of angles of incidence are :

$$|\theta| < 35.3° \tag{19}$$

The ranges of wavelengths are :

$$2.56 \ u^2/(m^2/sec^2) < \lambda/(m) < 3.83 \ u^2/(m^2/sec^2) \tag{20}$$

The wavelengths of blocked waves for specific currents are shown in Table 6. The smaller wavelength for each value of current corresponds to a wave at normal incidence while the larger value represents the wavelength when the angle of incidence is 35.3°. These two extremes include the range of wavelengths observed on the surface in the presence of a given blocking current.

Table 6. Waves blocked as they appear in the counter current

Current Speed (m sec^{-1})	Range of Wavelengths Blocked (m)
0.25	0.16 - 0.24
0.50	0.64 - 0.96
0.75	1.44 - 2.15
1.00	2.56 - 3.83
1.25	4.00 - 5.98
1.50	5.76 - 8.62

For representative values of u in the measured range of currents encountered in the field, that is, 0.75 to 1.25 msec^{-1} (20) yields wavelengths for the blocked waves between 1.44 and 4.00 m for normal incidence and values no larger than 5.98 m for a 35.3° incident angle. This wavelength range is not at odds with qualitative estimates based on visual inspection of the sea surface photos taken at Asia Rip. It remains to be demonstrated whether this is quantitatively consistent with observations. For purposes of argument we will consider the origin of the waves to be earlier winds.

We have looked at the wind records from the USNS HAYES for the 24 hours prcceeding the time when the photos were taken. Up to 5 hours before the photos were taken the wind was blowing with a westerly component at about 5 m/sec. Toward the west, fetch is limited only by the distant continental U.S.which is more than 180 km away depending on the precise direction that the light winds traveled. This leads to a fully developed sea. The wind veered clockwise coming from the north at τ = -5 hr., then gradually moving to ENE during the observations. A north-south line marks the boundary between winds from the west (and so before τ = -5 hr.), and winds of essentially infinite fetch, but possibly limited duration (after τ = -5hr.). Thus, the waves that arrive at the observation site that originated in the west were primarily generated as much as five hours previously.

We will consider 5 m/sec as the wind speed blowing from the west for a long time before τ = -5 hr. Such a wind will produce mostly waves having periods shorter than $2\pi c/g$ = 3.2 sec. So at the observation site the majority of waves impining from the west have periods less than 3.2 sec and speeds less than 5 m/sec. Equation (18) indicates that 3.2 sec period waves will be blocked by currents of about 1.3 m/sec. Considering the uncertainties, this is not inconsistent with the measured currents.

CONCLUSION

To summarize our findings, we have observed the striking contrast in ocean surface waves that characterizes a rip. The excess intensity of short waves coincides with the lee side of a bathymetric feature (Phelps Bank). The surface wave effect existed at a time when currents were directed toward the west, where the water is deeper, and when previously generated sea runs toward the east. We have examined some physical mechanisms which may account for this phenomena and find that wave blocking by current is the most likely. We believe that these observations provide an important clue in interpreting one aspect of SAR imagery, namely surface wave expression of bathymetry in a shallow sea.

Needless to say, the oceanographic situation is much more complex than this simple kinematic explanation would imply, but until more data are available for interpretation, we consider this the most simple and likely explanation that is consistent with the measurements that we have.

ACKNOWLEDGMENT

The authors express their appreciation to Dr· Jack A.C. Kaiser for providing most of the meteorological, hydrographic and navigational records used in preparation of this report. We also thank the personnel of Reconnaissance Squadron VFP-306 for their expert aerial photography. We also acknowledge helpful discussions with Dr. Davidson Chen and Dr. Gaspar Valenzuela on wave-current interactions and interpretation of radar imagery of the ocean surface.

REFERENCES

Alpers, W.R., Ross, D.B., and Rufenach, C.L., (1981). On the detectability of ocean surface waves by real and synthetic aperture radar, J. Geophys. Res. 86: No. C7, 6481-6498.

Brown, W.E., Jr., Elachi, C., and Thompson, T.W., (1973). Oceanographic observation with imaging radar, Union Radio Sci. Int. Fall Meeting, Boulder, Colo.

Brown, W.E., Jr., Elachi, C., and Thompson, T.W., (1976). Radar imaging of ocean surface patterns, J. Geophys. Res. 81: No. 15, 2657-2667.

Chen, D.T., (1982). Surface effects due to subsurface processes: a survey, NRL Memorandum Report 4727, January 15, 1-40.

DeLoor, G.P., and Brunsveld van Hulten, H.W., (1978). Microwave measurements over the North Sea, Boundary-Layer Meteorology, 13: 119-131.

DeLoor, G.P., (1981). The observation of tidal patterns, currents and bathymetry with SLAR imagery of the sea, IEEE Jour. of Oceanic Eng. OE-6: No. 4, 124-129.

Johnson, J.W., (1947). The refraction of surface waves by currents, Trans. Am. Geophys. Union 28: No. 6, 867-874.

Jonsson, I.G., Skougaard, C., and Wang, J.D., (1970). Interactions between waves and currents, Proc. 12th Coastal Eng. Conf., Am. Soc. Civil Engineers, Vol. 1: 489-507, New York.

Kasischke, E.S., Schuchman, R.A., and Lynden, J.D., (1980). Detection of bathymetric features using SEASAT synthetic aperture radar-a feasibility study, Environ. Res. Inst. of Mich., Report 135900-2-F.

Larson, T.R., and Wright, J.W., (1974). Imaging ocean current gradients with synthetic aperture radar, Union Radio Sci. Int. Fall Meeting, Boulder, Colo.

Longuet-Higgins, M.S., and Stewart, R.W., (1960). Changes in the form of short gravity waves on long waves and tidal currents, J. Fluid Mech. 8: 565-583.

Longuet-Higgins, M.S., and Stewart, R.W., (1961). The changes in amplitude of short gravity waves on steady non-uniform currents, J. Fluid Mech. 10: 529-549.

McLeish, W., Swift, D.J.P., Long, R.B., Ross, D., and Merrill, G., (1981). Ocean surface patterns above sea-floor bedforms as recorded by radar, Southern Bight of North Sea, Marine Geology, 43: M1-M8.

Moskowitz, L.I., (1973). The feasibility of ocean current mapping via synthetic aperture radar methods, Amer. Soc. of Photogramm. Fall Convention, Lake Buena Vista, Fla.

Phillips, O.M., (1981). The structure of short gravity waves on the ocean surface, in Spaceborne Synthetic Aperture Radar for Oceanography, Beal, R.C., DeLeonibus, P.S., and Katz, I., editors, Johns Hopkins Univ. Press, Baltimore, Md.

Taylor, G.I., (1955). The action of a surface current used as a breakwater, Proc. Roy. Soc. Lond. A231: 466-478.

Unna, P.J.H., (1942). Waves and tidal streams, Nature, 149: No. 3773, 219-220.

Ursell, F., (1960). Steady wave patterns on a non-uniform steady fluid flow, J. Fluid Mech. 9: 333-346.

Valenzuela, G.R., (1978). Theories for the interaction of electromagnetic and oceanic waves, Boundary-Layer Meteorology, 13: 61-85.

Valenzuela, G.R., (1981). A remote sensing experiment in the Nantucket Shoals (SEBEX), IUCRM Symposium on "Wave dynamics and radio probing of the ocean surface," Miami Beach, Fla.; submitted for publication in the proceedings, Plenum Press.

WAVE-CURRENT INTERACTIONS: A POWERFUL MECHANISM FOR AN ALTERATION
OF THE WAVES ON THE SEA SURFACE BY SUBSURFACE BATHYMETRY

JAMES M. WITTING
U. S. Naval Research Laboratory, Washington, DC 20375 (U. S. A.)

ABSTRACT

   Radar observations of the ocean surface can image features of
the floor of shelf seas. The radars, however, respond to water
waves that never feel the bottom directly, because they are too
short to do so. Wave-current interactions are a mechanism that
can allow a coupling between the short waves sensed by the radars
to a nonuniform current tied to the bathymetry. Wave-blocking,
which can occur when a wind-wave field propagates into a spatially
increasing current, is especially effective in altering the
properties of the wave field. This paper applies a new numerical
model of water wave evolution, the unified waves model, to the
problem of wave blocking. The model is fully nonlinear. Results
show that linear wave theory adequately predicts the value of the
current that blocks a wave of given frequency, and the evolution
of wavelengths. The model sees no evidence of significant
reflected waves nor much disturbance beyond the blocking point.
The waves simply pile up and break.

1. INTRODUCTION

   Over the last few years we have learned to appreciate that
Synthetic Aperture Radars (SAR's) can image important surface and
subsurface features of the ocean, at least sometimes. Among these
features are the wakes of large ships, large-scale ocean waves,
internal waves, and prominent bathymetric structures. Under the
usual operating conditions, however, the SAR's electromagnetic
signal senses only small-scale features at or near the Bragg
scattering wavelengths (typically tens of centimeters), or broken
portions of breaking waves. The features that the SAR's sometimes
image have scales of tens to hundreds of meters. A fundamental
problem is to determine how the radar scatterers on the ocean
surface are modified by currents and by other waves in such a way
as to permit SAR's to image a large-scale scene.

Without fundamental understanding of how large-scale ocean features generate or modify the small-scale radar scatterers, it may be difficult to identify the source of prominent features in SAR imagery, and well-nigh impossible to forecast whether a known large-scale feature will be imaged. The research described here addresses one aspect of scale-coupling, namely wave-current interactions, particularly wave blocking, a mechanism that some researchers think is responsible for high-constrast imagery of continental shelf bathymetry by SAR's (DeLoor, et al., 1978; DeLoor, 1981; Gordon, et al., 1983; Gordon, et al., this volume). The research uses a newly developed model that I call the "unified waves model" to gain insight into the physics of wave-current interactions.

This paper is organized as follows: Section 2 gives a brief description of the unified waves model (for a fuller account, see Witting, 1982). Certain additional capabilities have been added to the model that are necessary to simulate the scene observed by Gordon, et al. These are also summarized here (for a fuller account, see Witting, 1983). Section 3 reports aspects of the propagation of waves traveling directly into a current of increasing magnitude. The analysis presented by Gordon, et al. indicates that, by the time waves are blocked by ocean currents, they travel almost directly into the current, irrespective of the direction traveled earlier. Thus, the dynamics of the into-the-current situation is highly relevant to waves at the near-blocking condition, where the validity and completeness of the well-known analytical theories of Longuet-Higgins and Stewart (1960,1961) and other investigators is not proved. The unified waves model allows for fully nonlinear waves and for reflected waves, and so might provide insight that the analytical theories do not yield. Section 4 discusses the results and their limitations.

2. UNIFIED WAVES MODEL

The unified waves model uses finite-difference techniques to calculate the properties of evolving nonlinear water waves in one horizontal dimension. It is now capable of treating irrotational and constant-vorticity irrotational waves in channels of gradually varying depth and breadth, with ambient steady currents present, and with forced or free waves. Model results have been compared with results from laboratory experiments and with those from other numerical and analytical theories. To date, the comparisons with theory show the model to be very accurate; some features seen in experiments are seen in the unified waves model, but not in other models. The following factors are responsible for this development:

1. The model uses exact prognostic equations in conservation form. A derivation of the exact momentum equation used is given in a report by Witting and McDonald(1982). It appears not to have been used in wave modeling before now.

2. Higher order expansions than used before connect the velocity variables that appear in the governing equations. This allows the model a) to incorporate long wave theory exactly, b) to include both shallow water and essentially deep water waves in the same model, and c) to represent fairly long nonlinear waves to one order better than Boussinesq. For this reason the model is called "unified." Figure 1 connects this model with other water wave theories.

3. The model employs a numerical method, pure leapfrog, that gives no unwanted numerical diffusion. The time-stepping is simple enough to analyze in some detail and to permit running efficiently on vector computers.

4. The model can take a time step equal to a space step (in nondimensional units in which the linear long wave speed is unity). This permits efficient machine computations, unlike methods that have been developed for the Korteweg-deVries equation (see Vliegenthart, 1971, for discussion). Moreover, this procedure removes any spurious numerical dispersion at the lowest order.

5. Finally, the diagnostic equations are cast in a form such that only tridiagonal matrix equations need to be solved. A very fast, fully vectorized algorithm is then used to invert the matrices.

Some recent capabilities have been added to the unified waves model for wave-blocking studies. These include: 1) comuputer implementation of the basic analysis to include channels of gradually varying depth, 2) provision of a readily analyzable wavemaker to generate a situation on the computer that resembles a wind-generated sea, and 3) the ability to send a steady flow through the computational channel. Witting(1983) describes these additional capabilities, and shows examples of each. At the time of this writing, not all of the additional capabilities have been consolodated into a simulation of the oceanic scene described by Gordon, et al. The third new capability, however, when used in a

ORDER OF VALIDITY OF VARIOUS THEORIES

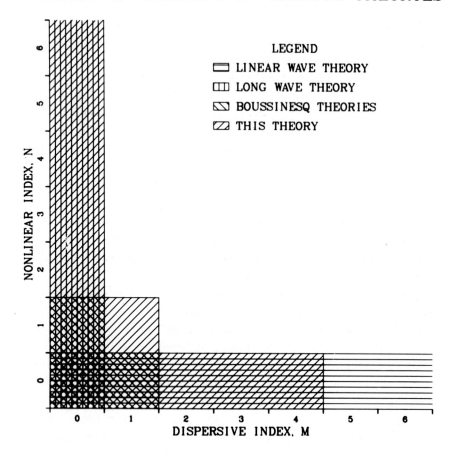

Figure 1. The formal accuracies of various water wave theories. The dispersive index M is the order of an expansion parameter defined by the square of D/L, where D is the water depth and L is a length scale associated with a wave (wavelength or solitary wave thickness, depending on the problem). Linear water wave theory goes to all orders of M. The nonlinear index, N, is the order of an expansion parameter defined by a/D, where a is the wave amplitude. Long water wave theory goes to all orders of N. Boussinesq theories retain only the lowest order dispersive terms and usually only the lowest order nonlinear terms. The unified waves model incorporates long wave theory, goes to fourth order in dispersion, and, for situations where dispersive and nonlinear effects are both important, goes one order beyond Boussinesq.

channel of variable breadth, permits the model to send a nonuniform
steady current through the computational domain, which is a key to
studying the fate of waves encountering nonuniform currents.

Running times for the unified waves model on the Texas
Instruments Advanced Scientific Computer, which is an early
Class VI machine, are approximately 20 msec. per time step for a
computational domain containing 1600 grid points. A run to
t = 150 with space and time steps each 1/8 takes about 24 sec.
About half of the running time is spent in collecting data for
detailed analysis, and can be eliminated, if desired. In
summary, computer costs to run the model are very small.

3. WAVE BLOCKING

This section is a mostly graphic display of what happens to
waves that propagate against a current whose magnitude increases as
seen by the waves. The unified waves model sets up the nonuniform
current by running a steady flow through a channel of variable
breadth. Figure 2 shows the channel geometry, surface elevation,
and horizontal component of the velocity of the undisturbed flow.
Because of Bernoulli's law, any variable velocity must be
accompanied by a change of surface elevation (this need not be the
situation if sidewalls are permeable). Consequently, the steady
component of surface elevation is not quite horizontal. The sign
convention is positive, left to right. The velocity is negative,
indicating flow from right to left, speeds decreasing from right
to left as the fluid encounters a widening channel.

Initial conditions are imposed that add to the steady flow a
wave packet that is designed to move from left to right, that is,
against an increasing current. Linear water wave theory is used
to design the relation between elevation and surface velocity to
limit the scene to right-going waves (the calculations, however,
permit waves traveling toward the left, and nonlinear effects can
lead to weak left-going waves). A fairly narrow wave packet with a
fixed carrier wave number of unity forms the disturbance. Figure 3
portrays the evolution of the packet. Initially, (the bottom trace
of the figure), the carrier wave number is unity, in nondimensional
units where the speed of long linear waves at the left side of the
computational channel is unity. Thus, the initial wave is intermediate between being a deep water wave and a shallow water wave
(the water depths and apparent surface scales observed by Gordon,
et al. are not scaled properly in these computations; their waves

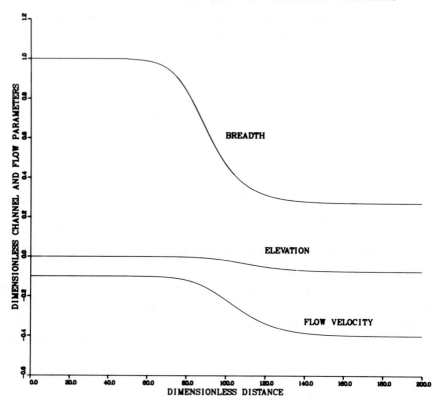

Figure 2. Ambient steady flow for calculations involving the
phenomenon of wave-blocking. A four-fold decrease in the magnitude
of a steady nonuniform current occurs smoothly as it flows from the
right side of the computational channel (x = 200) to the left side
(x = 0). The channel geometry that produces the nonuniform current
is that of a converging channel in the wave direction (left-to-
right), and of a diverging channel in the flow direction (right-to-
left). The calculations for the ambient mean water surface
elevation and fate of an initial wave packet subsequently described
are made with the unified waves model with space and time step
sizes each one-eigth the depth, in nondimensional units where the
speed of a long linear wave at zero elevation is unity.

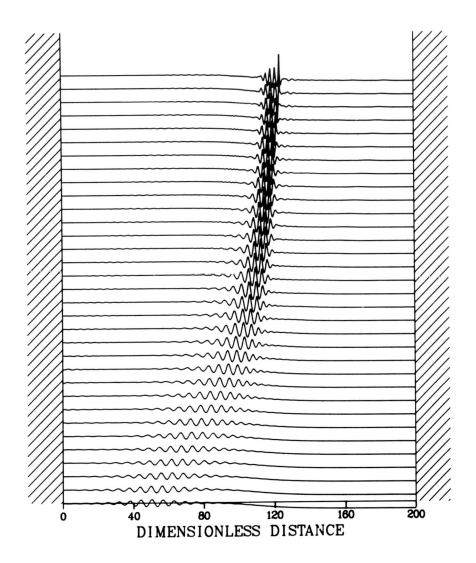

DIMENSIONLESS DISTANCE

Figure 3. Wave blockina by steady, nonuniform, adverse currents.
The figure shows 33 profiles of the water elevation at equally
spaced time intervals, starting at the bottom. The time intervals
are set so that a feature that moves with the speed of a long
linear wave would be aligned with a slope of unity in the figure.
The initial wave packet decelerates, its waves becoming shorter and
higher as it moves into the adverse current shown in Figure 2.
The highest waves in the packet would have broken before they
reach the top of the figure.

are in deep water. I select the initial wavelength so that when the waves are blocked, they are essentially deep water waves, but can still be represented with fidelity by the unified waves model.)

The other traces shown in Fig. 3 are "snapshots" of the surface elevations at uniformly-spaced later times. The interval between profiles is set so that a feature traveling at the long linear wave speed would be oriented at 45 degrees in the figure. In short, Fig. 3 is an x-t diagram with structure.

Figures 4-10 blow up the uppermost 21 traces of the 33 shown in Fig. 3. Before turning to them, note from Fig. 3 that there is no evidence of significant reflected waves. The kinematics permits these, though most analytic formulations limit themselves to ignoring reflected waves. It appears that the neglect of reflected waves is satisfactory, at least for the example shown.

Linear theory predicts that waves having the frequency of the initial wave packet should be blocked at x = 119. The location of the piling up of waves in Figure 3 is near x = 119, and wave amplitudes are very small further to the right.

The details in Figs. 4-10 show the wavelengths decreasing and the amplitudes increasing as the packet progresses into the current. Some spot checks on wavelengths show that they too are in accordance with linear theory. The agreement between the wavelengths and blocking locations observed by the unified waves model and linear water wave theory is not surprising. Nonlinear effects on the speed of waves are at most 19 percent, and the wave speed is the only variable affecting wavelengths and blocking location.

The final six profiles are beyond the conditions taken by actual water waves (Figs. 9-10). The amplitude and maximum slope of a progressive wave just starting to break lie near the theoretical values of wave-height to wavelength ratio 0.141 and 30 degrees. Figure 11 expands the region of the highest wave of the 7th and 6th uppermost profile of Fig. 3, with no vertical exaggeration. These highest waves happen to be located at about the same grid points, but they are not the same wave (the phases are moving at good speed even as the group is stopped; the crest of the lower profile of Fig. 11 was one crest behind that of the upper profile during the upper snapshot). The upper profile does not exceed a bound of unbroken waves. The lower profile exceeds both the amplitude and slope of a breaking wave.

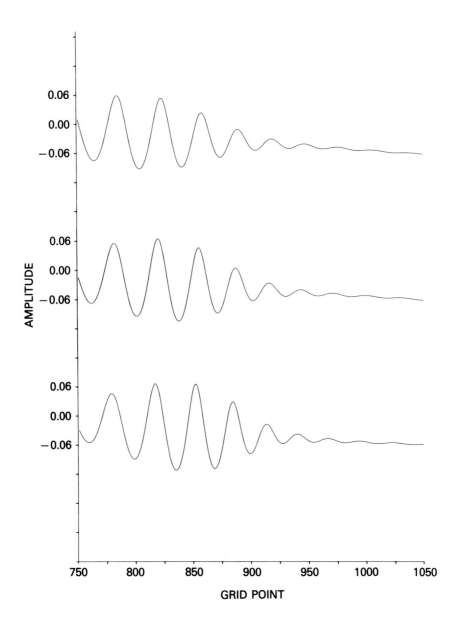

Figure 4. Expanded view of the 13th profile from the bottom of
Fig. 3 (at top here), the 14th profile, and the 15th profile (at
bottom here). There are eight grid points per x-interval, so the
left side here corresponds to x = 93.75, and the right side here to
x = 131.25 of Fig. 3.

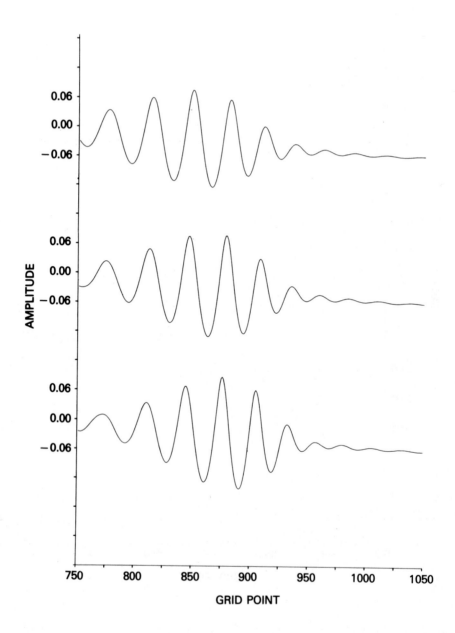

Figure 5.   The 16th (top), 17th, and 18th profile from the bottom
of Fig. 3.   See caption of Fig. 4.

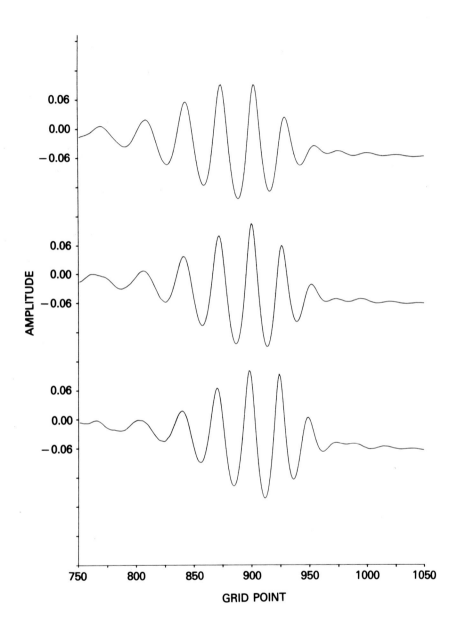

Figure 6.   The 19th (top), 20th, and 21st profile from the bottom
of Fig. 3.   See caption of Fig. 4.

198

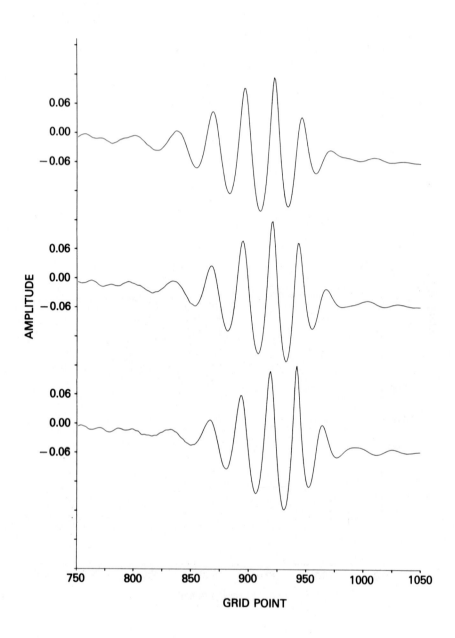

Figure 7.   The 22nd (top), 23rd, and 24th profile from the bottom
of Fig. 3.   See caption of Fig. 4.

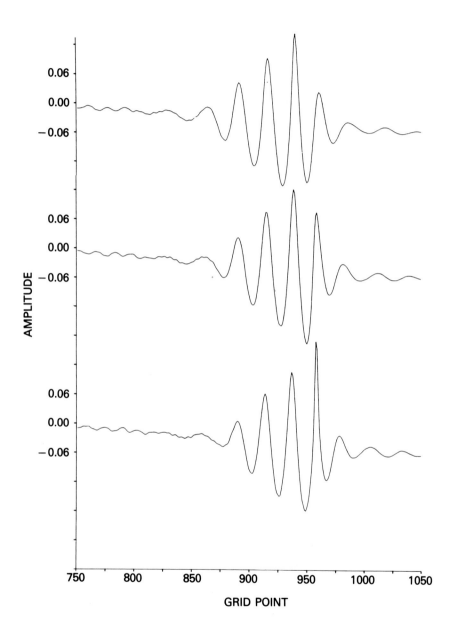

Figure 8.   The 25th (top), 26th, and 27th profile from the bottom
of Fig. 3.   See caption of Fig. 4.

200

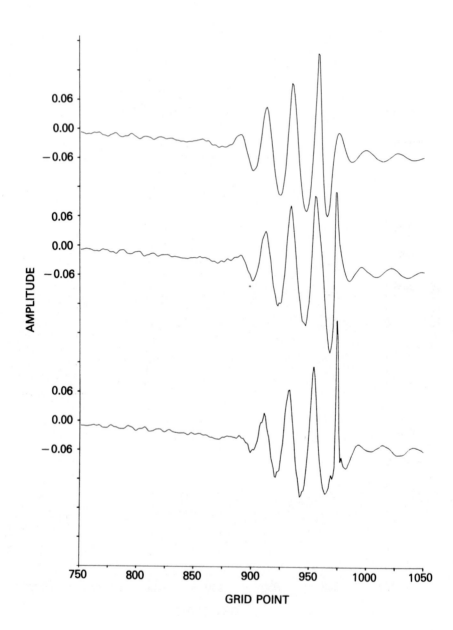

Figure 9.   The 28th (top), 29th, and 30th profile from the bottom
of Fig. 3.   See caption of Fig. 4.

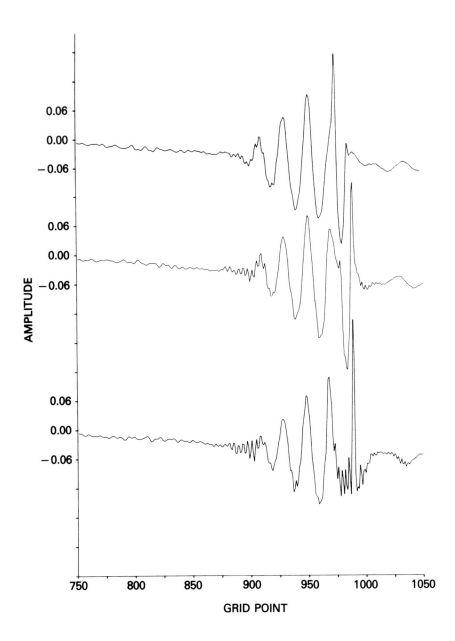

Figure 10.   The 31st (top), 32nd, and 33rd profile from the bottom
of Fig. 3.   See caption of Fig. 4.

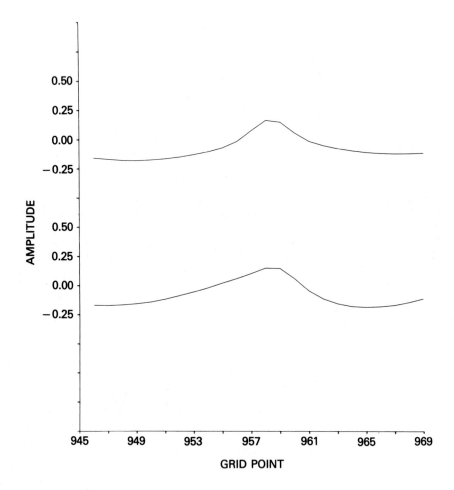

Figure 11. An expanded view of the 27th (top) and 28th profile from the bottom of Fig. 3. The horizontal and vertical scales are the same, so that ratios of wave height to depth and slopes are viewed without distortion. The 28th profile has a wave height and slope exceeding that of a spilling breaking wave.

4. DISCUSSION

This is essentially a progress report of the application of the unified waves model to wave-current interactions in general, and to wave blocking in particular. The setting is a steady nonuniform flow into which a wave packet propagates. The waves become shorter and higher. Some of the increase of amplitude is due to the change of channel breadth (less than half, if Green's law, 1838, applies). The rest of the change of amplitude results from wave-current interaction.

The model gives changes of wavelength and position of blocking that agree well with what linear water wave theory predicts. There is no evidence of significant reflected waves, nor of significant evanescent waves ahead of the blocking position. The basic picture that the unified waves model draws is that of breaking waves at the place that the group speed of the waves is matched by the current, and little else.

REFERENCES

DeLoor, G.P. and Brunsveld van Hulten, H.W., 1978. Microwave measurements over the North Sea. Boundary-Layer Meteorology, 13: 119-131.

DeLoor, G.P., 1981. The observation of tidal patterns, currents and bathymetry with SLAR imagery of the sea. IEEE Jour. of Oceanic Eng., OE-6, No. 4: 124-129.

Gordon, C., Greenewalt, D., and Witting, J., 1983. Asia rip: Surface wave expression of bathymetry. NRL Memorandum Report 5027, 40pp.

Gordon, C., Greenewalt, D., and Witting, J. Surface-wave expression of bathymetry over a sand ridge. This volume.

Green, G., 1838. On the motion of waves in a variable canal of small depth and width. Camb. Phil. Trans., 6: 457-462.

Longuet-Higgins, M.S. and Stewart, R.W., 1960. Changes in the form of short gravity waves on long waves and tidal currents. J. Fluid Mech., 8: 565-583.

Longuet-Higgins, M.S. and Stewart, R.W., 1961. The changes in in amplitude of short gravity waves on steady non-uniform currents. J. Fluid Mech., 10: 529-549.

Vliegenthart, A.C., 1971. On finite-difference methods for the Korteweg-deVries equation. J. Eng. Math., 5: 137-155.

Witting, J.M., 1982. A unified model for the evolution of nonlinear water waves. NRL Memorandum Report 5001, 64 pp. Submitted for publication in J. Comp. Phys.

Witting, J.M., 1983. Additional capabilities of the unified waves model. NRL Memorandum Report 5194, 52 pp.

Witting, J.M. and McDonald, B.E., 1982. A conservation-of-velocity law for inviscid fluids. NRL Memorandum Report 4977, 12 pp. Submitted for publication in J. Comp. Phys.

ACKNOWLEDGEMENT

This work is supported by the Coastal Sciences Group, Ocean Sciences Division, Office of Naval Research.

REMOTE SENSING OF OIL SLICK BEHAVIOUR

P. P. G. DYKE
Mathematics and Computer Studies Dept., Sunderland Polytechnic,
Sunderland. SR1 3SD. U.K.

ABSTRACT
   This short contribution highlights the role remote sensing has
played in our understanding of the behaviour of sea surface oil.
In particular, it points out the importance of Langmuir Circulations an effect that received derisory attention before the
advent of optical and infra-red aerial-photography. Finally, some
attempt is made to forecast the future role of remote sensing in
oil slick monitoring.

INTRODUCTION
   The principal aim of this paper is to assess the impact that
remote sensing has had upon our understanding of the behaviour of
oil slicks in open sea.
   Remote sensing has been around since the Second World War, and
as a "revolution" seems to have been largely overlooked by the
general public. Essentially, remote sensing is as old as the
camera, it is the term "remote sensing" that is new, and this
phrase has been coined in response to the bewildering array of new
instruments that have appeared in the last decade or so. Every
remote sensing device is a transmitter and receiver of signals,
the most useful devices for oil slick examination use the optical
to radio-wave section of the electromagnetic spectrum. The atmosphere is too opaque to shorter wavelengths, although the use of
ultra-violet sensors has been reported, to utilise oil slick
luminescence.
   The main techniques used to study sea going patches of surface
or near surface oil are (i) optical photography and (ii) infra-
red line scan together with employment of satellites. Acoustic
techniques have been used, but are mainly reserved for beneath-the-
sea navigation. The use of infra-red devices is particularly
useful in the detection of illegal dumping of waste oil which
usually takes place at night or in low visibility conditions.

First of all, let us examine the role of aerial photography.
The advantages of this technique are in its simplicity and in the
ease of interpretation of the results. The results of competent
aerial photography can be easily understood by the layman, an
important factor if pressure for new legislation is being applied.
Aerial photographs give a good overall impression of what is occurring, but often do not show detail. Only in the most favourable
circumstances does it reveal anything about what is happening under
the surface of the sea.

Early investigators found that shipborne sensors were inadequate
for oil slick monitoring. The oblique angle of observation and the
closeness of the ship's horizon made a large surface oil slick
difficult even to detect, let alone monitor. It was the Torrey
Canyon disaster of 1967 that first showed how useful IRLS could be.
In 1977, the Ekofisk blowout was monitored using 8-14 μm IRLS and a
50kW(H) polarised Q-band SLAR (Side Looking Airborne Radar) over-
flying a spill of 5 tonnes of Ekofisk crude (Parker & Cormack
(1979)). Parker & Cormack (1979) give an account of controlled
spills and detail how thicknesses of slicks can be measured using
the IRLS technique. Their paper is mainly concerned with the
comparison between IRLS and SLAR conclude that the former gives
better resolution and more information on slick thickness than the
latter.

The two factors that govern the usefulness of a remote sensing
device for oil slick examinations are (i) the penetration capability of the chosen radiation, and (ii) the variation of the
incoming signal, be it reflected or back-scattered, with parameters
of interest such as temperature, dissolved substances, suspended
matter etc., in the region of sea being studied. In view of these
factors, the infra-red line scan technique is used. IRLS successfully detects the differences in thickness of surface oil slicks
because the thicker part of the slick is warmer than its surroundings due to the adsorption of heat radiation from the sun. This
difference in temperature alters the character of the backscattered
radiation (that is the overall signal to noice ratio of the detect-
or) and shows up as a bright area on the IRLS photograph (see
Figure 1).

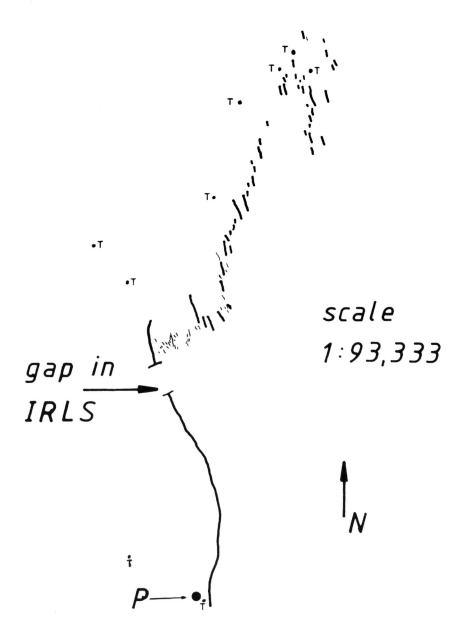

gap in
IRLS

scale
1:93,333

Fig. 1. An aerial photograph using an infra-red technique of oil
on the sea surface at 28th April, 1977. This has been traced from
two plates in Audunson (1978) so that the black streaks are oil
(they were white on black in the original). The dots 'T' are vessels monitoring the slick from the surface and 'P' is the Bravo
platform itself. Note the breakup into streaks.

As mentioned at the outset, the purpose of this paper is to help to understand the behaviour of the oil slicks, not in a detailed appraisal of particular remote sensing devices. So, although it is recognised that interpretation and data analysis procedures are extremely important, these will receive no attention in what follows. It is felt safe to do this as the majority of the facts presented have come though optical and infra-red photography, about which there are few arguments nowadays.

Oil Slick Constituents

An important question to address is what constituents of oil slicks are harmful, and to what degree does the sea's dynamics separate or comingle these constituents? Of course, detailed appraisal of the various crude oils is within the realm of the maring chemist, however, physical oceanographers must be aware of some of the vast differences in density, viscosity etc., of say, bitumen and light crude.

It is perhaps useful here to distinguish between tanker accidents and oil production platform blowouts. In general, it is tanker accidents that provide the greatest risk to the marine ecosystem. Oil like Kuwait crude is heavy and difficult to deal with. Blowouts on offshore platforms usually give less oil, and the oil itself is much lighter with a tendency to disperse and evaporate (e.g. The Bravo (Ekofisk) blowout of 1977, Audunson (1978)). It is much more important, therefore, to understand how the heavier oils of the Middle East behave in the sea.

When oil is discharged into the sea several things happen simultaneously. It separates by density, lighter crude floating and heavier sinking, sometimes residing at an intermediate density interface such as the thermocline (there was evidence of this in the photographic data of the IXTOC I blowout in the Gulf of Mexico where oil was being transported towards Texas approximately 50m below the sea surface, probably at a pycnocline - Attwood (1980)). Some of the heaviest oils go into tar balls which separate out from a slick. The precise manner in which this is achieved is not well understood but it is promoted by the use of detergents. After a while, the oil and water may form a "chocolate mousse", an emulsion containing over 50% water and remaining at the sea surface as a film over 1mm thick. Smith (1968) reported such emulsions moving from the Torrey Canyon. The oil and water also interact

chemically in a manner dependent on factors such as the temperature
and salinity of the sea, and the composition of the crude. The
dynamic interaction of oil and water is better but certainly not
completely, understood and is the subject of the next section.

Oil Slick Movement

As offshore oil was discovered, larger and larger oil tankers
were being built, and stricter legislation was being drafted. As
the risk of catastrophic oil spills increased, so the insurance
companies needed more calculations and data upon which to base
their levies. This lead to publications such as Dietzel, Glass &
Van Kleef (1976). In the body of this report had to be an answer
to a very difficult question, namely:- if a given oil well had a
blowout offshore and a given amount of a specified crude oil
escaped into the sea at a given location then how much of it would
reach land in a given direction a given distance away. Of course,
a great deal of stochastic modelling had to be done, but the under-
lying dynamical assumption was that the sea surface borne
contaminant moved at 3.4% of the wind speed in a direction 15° to
the right of it (in the northern hemisphere). Clearly, however,
the emphasis in such models is not on the dynamic aspects.

The agencies that cause water borne oil to move include diffu-
sive spreading, wind and wave action, sea currents (including
tides, convection currents and wind driven currents) and Langmuir
Circulations. Agencies that cause the oil to break up include the
formation of emulsions, chemical reactions, variations in oil
density and therefore, in particle sinking rates, biological
action, evaporation and precipitation, and atmospheric oxidation.
There is no doubt that remote sensing has given a clearer insight
into some of these processes.

It is not possible to separate these agencies from one another
since they all contribute to oil slick break up. It should be
pointed out that even with all these agencies in operation, oil
slick dispersion seldom occurs naturally. Most large oil slicks
need man's help in clean-up procedures, the most perferable being
collection by booms or polystyrene, or burning. Chemical dispersion is used as a last resort. Note, however, that the chocolate
mousse emulsion mentioned earlier will not burn, there is too
much water in it.

The next section is devoted to a mechanism mentioned above that
has not previously received the attention it deserves as a mechanism for moving surface contaminants, namely Langmuir Circulations.

Langmuir Circulations

Langmuir Circulations are helical roll vortices in the surface
layers of the ocean whose existence has been dramatically established in recent years by the behaviour of sea surface oil and its
subsequent detection by remote sensing. The two photographs,

Plate 1. An aerial photograph of a surface patch of oil. Note the
breaking up into streaks, and the windrows (about 100m apart).
(Photograph courtesy of Warren Springs Laboratory and published
with their permission).

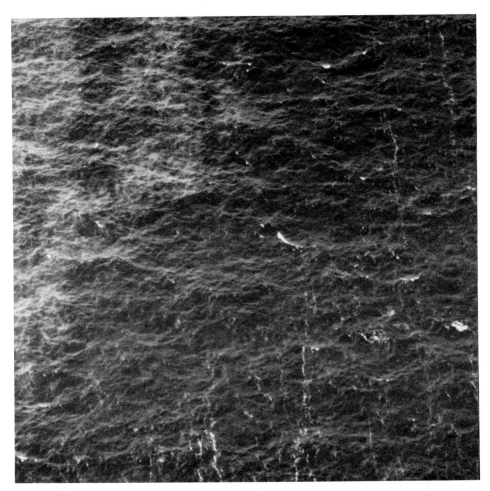

Plate 2. An aerial photograph of a small slick being eroded by waves. Note the streaky appearance, possibly caused by Langmuir Circulations. (Photograph courtesy of Warren Springs Laboratory and published with their permission).

Plates 1 and 2, show the oil on the sea surface tending to split into streaks in a manner dictated by the hydrodynamics of Langmuir Circulation. It is perhaps wise to consider this phenomenon more carefully and study the consequences.

Langmuir Circulations (named from the seminal paper Langmuir (1938)) have long been recognised as responsible for the familiar lines of foam about 5 metres apart that occur at right angles to the wave crest lines in inland waters (e.g. lakes and reservoirs). Remote sensing has shown similar "windrows" at sea. This time,

however, the spacing is of the order 100m which makes detection
at the sea surface itself (for example from a boat) difficult.
Mathematical theories of Langmuir Circulations have appeared in
the last decade or so, and from an examination of them it seems
that there is no obvious explanation for these circulations.
Several different theories have been proposed, and the interested
reader is directed towards the recent excellent review of Leibovich
(1983). If a coherent spatial structure exists in the surface
waves, then a spatially periodic Stokes drift will be induced.
This periodic drift will produce a torque due to horizontal
variations in the vortex force that will directly drive the roll
motions. However, an instability mechanism in the undirectional
current will also produce Langmuir Circulations even without any
coherent surface waves. In this latter case, the vortex force is
balanced by a vertical pressure gradient. Direct wind-driven
Langmuir Circulations can also take place (Craig and Leibovich
(1976)).

Some of the biological and ecological consequences of Langmuir
Circulations are highlighted in Dyke & Barstow (1983). The
important feature of the circulations is the juxtaposition of
each vortex, producing lines of downwelling and upwelling zones
(see Figure 2). The magnitude of the upwelling and downwelling
in the divergence and convergence zones (respectively) will govern
precisely what will remain near the surface and what will sink
beneath it. Biologists have known for some time that convergence
zones cause enhanced production by concentrating the plankton into
lines, but the same convergence zones can also help to concentrate
surface oil into thicknesses that are potentially harmful to the
environment. Dyke & Barstow (1983) report details of many observations that support the conjecture that the convergence zones act
as a source of negative diffusion. This is particularly dangerous
as hitherto non-toxic levels of pollutant could be concentrated up,
to become toxic.

All of the research into Langmuir Circulations mentioned above
owes a great deal to the techniques of remote sensing. The structure of the circulations is such that only by detection from a
distance can the overall patterns emerge. This is shown graphically by the plates here and those in Dyke & Barstow (1983).

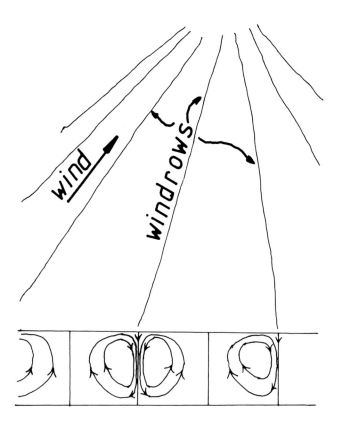

Fig. 2. An idealised representation of Langmuir Circulation.

Future Uses of Remote Sensing

Remote sensing is ideally suited to monitoring the behaviour of
oil slicks, simply because oil slicks provide a wholly unpleasant
environment. The further one is away, the better! Perhaps the
greatest advance in the next few years will be in the use of satellite borne infra-red devices for oil slick monitoring. Already the
motion and general behaviour of the larger slicks, for example the
very unpleasant spill in the Middle East, a consequence of shallow
water production platform being damaged in the Iran-Iraq conflict,
can be tracked by satellite. With improvements in microprocessor
controlled optics, it can be expected that this tracking ability
will soon be extended to cover slicks only a few square metres in
surface area. What is certain is that a geostationary satellite
with infra-red sensors trained on offshore loading and unloading
terminals would be a significant deterrant for those who persist
in illegally flushing out their tanks at sea.

As satellites become more common, it may be possible to undertake spatial modelling of oil slicks in much the same way as is being done in meteorology. This technique requires many samples, virtually a "film", of the particular slick's movements from which to deduce dynamics. This technique is furthest advanced in meteorology due to GARP (the Global Atmospheric Research Programme) but oceanographers are already examining the possibility of improving our knowledge of oceanic fronts and similar mesoscale phenomena. It must be stressed, however, that there are still technical problems to be overcome before the existing infra-red techniques can be used with confidence to deduce truly three dimensional behaviour at sea.

In closing, mention ought to be made of the work of Shearman (1980). He demonstrates the use of ionospheric backscatter whereby radio waves are refracted back to the desired area of sea by the ionosphere. This technique is suitable for surface wave detection and, since oil damps out waves, can therefore be applied here. However, the technique remains contraversial and the accuracy of results is still open to doubt.

Acknowledgement

The author would like to thank Dr. Stephen Barstow who carried out much of the preliminary work whilst working as a Research Assistant with him at Heriot-Watt University's Department of Offshore Engineering.

REFERENCES

Attwood, D. K. (ed), 1980. Preliminary Results from the September
1979 RESEARCHER/PIERCE IXTOC I Cruise. U.S. Dept. Commerce
NOAA/RD/MP3, Boulder, Colorado, U.S.A.
Audunson, T., 1978. The Bravo Blowout. IKU Report No. 90,
Trondheim, Norway.
Craik, A. D. D. and Leibovich, S., 1976. A Rational Model for
Langmuir Circulations, J.Fluid.Mech., 73, 401-426.
Dietzel, G. F. L., Glass, A. W. and Van Kleef, P. J., 1976.
"Sliktrak. A computer simulator for the prediction of slick
movement by natural means, clean-up and potential damages
arising from oil spills originating from offshore oil well
blowouts. Its development and application to the North Sea".
Shell Internationale Petroleum, Maatshappij, B. V., The Hague,
The Netherlands, Report EP-47436, December, 1976.
Dyke, P. P. G. and Barstow, S. F., 1983. The importance of
Langmuir Circulations to the Ecology of the mixed layer pp
486-497 in "North Sea Dynamics" eds. Sundermann and Lenz,
Springer-Verlag.
Langmuir, I., 1938. Surface motion of water induced by wind,
Science 87, 119-123.
Leibovich, S., 1983. The form and dynamics of Langmuir Circula-
tions. Ann.Rev.Fl.Mech 15, 391-427.
Parker, H.D. & Cormack, D., 1979. Evaluation of infra-red line
scans (IRLS) and side-looking airborne radar (SLAR) over control-
led oil spills in the North Sea. Warren Springs Laboratory
Report LR315(OP), Stevenage, U.K. 25pp.
Shearman, E. D. L., 1980. Remote sensing of the sea surface by
decametric radar. Radio and Electronic Engineering 50, 611-
623.
Smith, J. E. (ed), 1968. "Torrey Canyon" pollution and marine life.
C.U.P.

AN INTERCOMPARISON OF GEOS-3 ALTIMETER AND GROUND TRUTH DATA
OFF THE NORWEGIAN COAST

Lygre, Asle

 Oceanographic Department, Continental Shelf Institute,
 P.O. Box 1883, 7001 Trondheim (Norway)

ABSTRACT

    An intercomparison has been carried out between wave heights
and wind speeds estimated from the GEOS-3 altimeter data off
the Norwegian coast. The main result of this study has been
that the absolute mean difference between GEOS-3 measurements
of significant wave height (HS) and corresponding waverider
measurements is 1.18 m. Meanwhile the wind measurements show a
corresponding difference of 2.43 m/s.
    In addition the tendency was towards overestimation of HS
by GEOS-3 compared to HS measured by the waveriders. We have
tried to account for this overestimation but no obvious
explanation has been found. However, we might suggest the
following as possibilities: spatial non-homogeneities in the
wave field caused by, for example, the coastal shielding
effect, instrument malfunction in the altimeter as the
satellite ground track passes from land to sea, possibly due to
the change in reflectivity. Surprisingly this systematic
overestimation of HS by GEOS-3 has not been reported in earlier
comparisons*, indicating that altimeters are possibly not
suitable for measuring waves in near coastal areas.
    To assess if and why there are inconsistencies between
satellite and waverider measurements off the Norwegian coast
further investigations must be carried out. Unfortunately, the
data collected by SEASAT in 1978 comprises only a few passes
over Norwegian waters. It was not possible to give a definite
answer to the question as to whether or not there exist positive wave gradients in a direction away from the coast.
However, the satellite measurements did show situations with
relatively large geographical variations in HS. In the Halten
area, increasing wave height with distance away from the coast
always corresponded to approaching frontal systems. The GEOS-3
passes giving data across frontal systems shows how powerful
satellite measurements can be in monitoring the sea state, but
before this can be achieved accurately, an algorithm for
correcting the effects of clouds and heavy precipitation is
required.

* Except for comparisons of HS recorded early in the GEOS-3
   mission (Fedor et al., 1979). The error was detected and we
   have assumed that this has been corrected for in the
   algorithm of concern.

INTRODUCTION

The GEOS-3 satellite was launched on April 9th 1975 from the Air Force Western Test Range. One of the purposes was to test the possibility of collecting geophysical data by means of a radar altimeter. Data was acquired from April 21st 1975 until December 8th 1978. The data used in this comparison was provided on tape by NOAA's satellite division in Washington DC. The tapes contained the corrected data set from the GEOS-3 mission and covered the North Atlantic area.

In this paper we will concentrate on the GEOS-3 wave measurements and compare them to measurements of significant wave height provided by three waverider buoys located at Utsira, Brent and Halten off the Norwegian coast - see fig. 1. We will also compare wind measurements at Utsira to wind speed deduced by the scattering of the radar return signal. Wave profiles in a direction away from the coast are also considered to see if buoy measurements near the coast line can be used to reflect the sea state further off the Norwegian coast. Two passes of the GEOS-3 satellite over meteorological frontal systems are also presented.

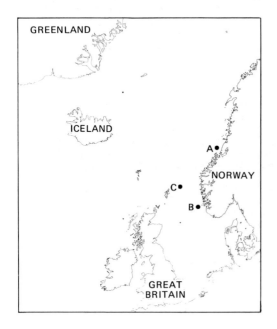

Fig. 1. Map showing the waverider locations at Halten (A), Brent (C) and Utsira (B).

THE COMPARISON OF GEOS-3 AND IN-SITU DATA

For those who are interested in a description of the altimeter and the GEOS-3 satellite we refer to the special GEOS-3
volume of Journal of Geophysical Research vol. 84, and Barrick
(1972).

When comparing GEOS-data to in-situ data one should be aware
of the following:

1) Wind and wave measurements obtained by GEOS-3 are in effect
   moving averages along the ground track. In this report HS is
   averaged over approximately 140 km.

2) In-situ measurements are based on time series at a fixed
   point on the sea surface.

3) There are space and time lags between the ground measurements
   and the GEOS-3 measurements.

Considering points 1 to 3 it is obvious that differences
between GEOS and in-situ measurements may occur due to for
example, effects of rapid changes in the surface weather
situation both in time and space. To a certain extent it is
possible to control this feature by an inspection of weather
maps to assure reasonable stationarity in the weather
situation.

In the following comparison the difference in time between
ground and satellite measurements seldom exceeds 2 hours and
the geographical distance between the waverider and the point
of nearest approach lies between 8 and 140 km. Most of the data
considered here is from February 1976, but data was also chosen
from other months in 1975 and 1976.

Figs. 2 to 4 show the ground tracks of the GEOS-satellite
during the selected passes in the vicinity of the waveriders.
Table 1 further gives the time and date of the passes. As is
readily seen there is a geographical spread in the passes near
the buoy positions. Sometimes it was difficult to pick out a

representative estimate of HS due to rapid variations in HS
along the track close to the waverider. However, as a rule the
nearest measured value was chosen, but there also arose
situations where it was necessary to average the HS estimates
along the track close to the waverider position and use this
mean value when comparing to the waverider data. This averaging
is in addition to the averaging performed by the computer
algorithm and will certainly cause additional smearing of the
estimate. In an attempt to minimize the effect of time lag
between ground measurements and satellite data, the buoy values
were interpolated linearly with time. Buoy measurements of HS
were available every third hour.

With regard to the comparison of wind speed at Utsira Lighthouse to wind estimates obtained by GEOS-3, the same procedure
as outlined above was used. The wind data at Utsira Lighthouse
was provided by the Meteorological Institute in Oslo.

In Fig. 5, scatter plots of GEOS-3 and waverider data are
shown in addition to a scatter plot of wind speed measured at
Utsira and by GEOS-3. Two lines of regression are fitted to the
data points, one of which is forced to run through the origin.
For the wind speed data and the Brent data only one line of
regression is shown in the scatter diagrams as the two lines
coincide.

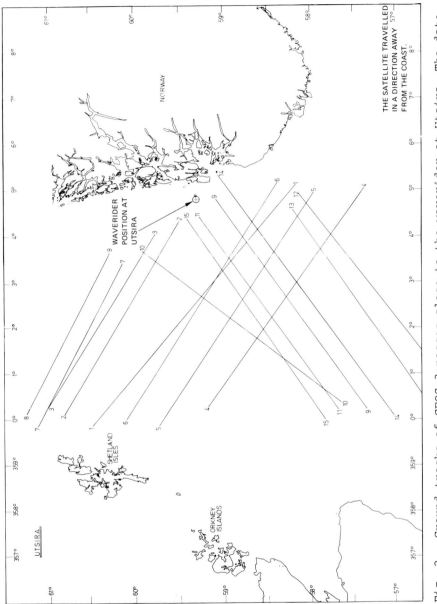

Fig. 2. Ground tracks of GEOS-3 passes close to the waverider at Utsira. The date and time of each pass is given in Table 1.

222

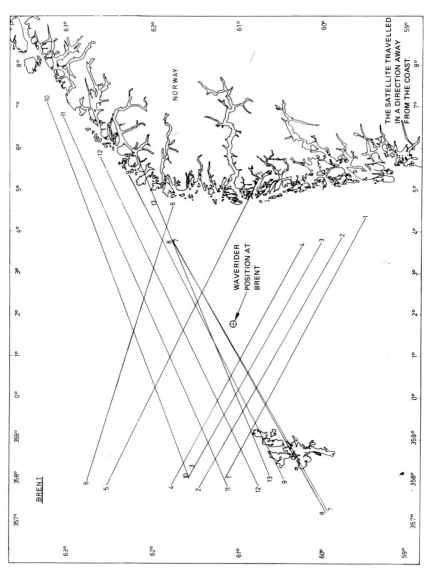

Fig. 3. Ground tracks of GEOS-3 passes close to the waverider at Brent. The date and time of each pass is given in Table 1.

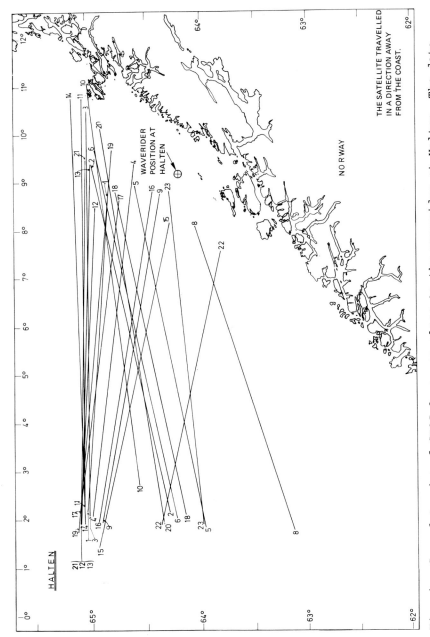

Fig. 4. Ground tracks of GEOS-3 passes close to the waverider at Halten. The date and time of each pass is given in Table 1.

HALTEN		
	DATE	TIME
1.	090276	8.13
2.	120276	7.30
3.	150276	8.26
4.	130276	8.55
5.	130276	7.16
6.	080276	6.49
7.	060575	22.07
8.	140276	7.02
9.	080276	8.28
10.	061175	20.09
11.	230475	21.56
12.	220475	22.10
13.	190276	9.07
14.	110276	8.11
15.	170276	9.36
16.	220276	10.03
17.	180276	9.22
18.	220276	8.24
19.	230276	9.48
20.	170276	7.57
21.	100276	7.59
22.	160276	9.50
23.	230276	8.10

BUOY POSITION:
N64^{o} 11' E9^{o} 8'

UTSIRA		
	DATE	TIME
1.	210276	11.56
2.	020676	23.43
3.	141075	20.44
4.	170775	10.45
5.	160276	10.29
6.	070276	11.21
7.	030276	9.40
8.	170276	11.15
9.	280775	6.30
10.	220276	6.47
11.	050776	22.25
12.	180276	6.05
13.	020875	6.57
14.	241075	16.42
15.	080276	5.11

BUOY POSITION:
N59^{o} 18' E4^{o} 48'

BRENT		
	DATE	TIME
1.	020676	23.42
2.	260276	12.22
3.	030276	9.40
4.	170276	11.16
5.	130276	10.34
6.	040276	9.26
7.	080776	23.20
8.	110276	6.07
9.	180776	00.28
10.	070776	23.34
11.	150276	6.49
12.	060276	5.40
13.	200276	7.16

BUOY POSITION:
N61^{o} 4' E1^{o} 48'

Table 1. The date and time of the GEOS-3 passes close to the waveriders at Utsira, Brent and Halten as shown on the maps in figs. 2 to 4. Each pass is numbered according to the numbers given on the trajectories in figs. 2 to 4.

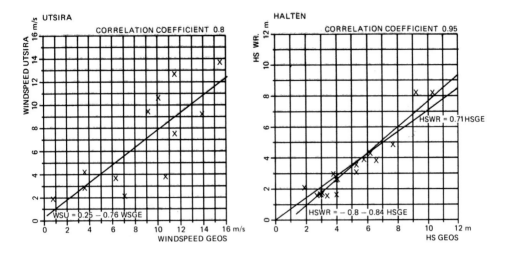

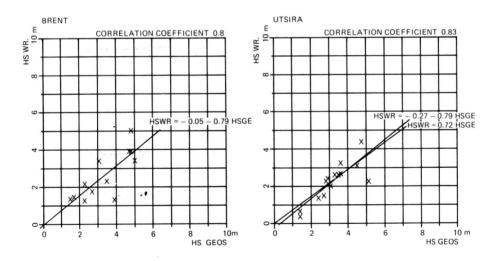

Fig. 5. In-situ measured wind speed and wave heights plotted against corresponding GEOS-3 measurements. For the Halten and Utsira wave data 2 lines of regression are specified one of which runs through the origin. The correlation coefficient is given on top of each scatter plot.
HSGE: HS from GEOS-3. HSWR: HS from waverider. WSU: Wind speed from Utsira Lighthouse. WSGE: Wind speed from GEOS-3.

	UTSIRA	BRENT	HALTEN	UTSIRA AND BRENT	TOTAL	C.L. PARSONS	L.S. FEDOR et al
ABSOLUTE MEAN DIFFERENCE $\frac{1}{N}\Sigma$ (HSGE − HSWR)	0.94 m	0.82 m	1.62 m	0.88 m	1.18 m		0.88 m
MEAN DIFFERENCE $\frac{1}{N}\Sigma$ (HSGE − HSWR)	0.94 m	0.73 m	1.60 m	0.84 m	1.14 m	0.24 m	0.19 m
STANDARD DEVIATION OF MEAN DIFFERENCE	0.61 m	0.75 m	0.71 m	0.69 m	0.79 m	0.53 m	1.02 m
CORRELATION COEFFICIENT	0.83	0.80	0.95	0.81	0.91		0.67
MEAN HS WAVERIDERS/BUOYS	2.27	2.58	3.36	2.42	2.80	high sea state	low sea state
PERCENTAGE OF GEOS DATA WITHIN ±0.5 m OF THE WR − DATA.	28.6 %	38.5 %	5.6 %	33.3 %	22.2 %		33 %
NUMBER OF PASSES (N)	14	13	18	27	45	19	27

Table 2. Summarizes the comparison parameters in this study and the comparisons carried out by Fedor et al. (1979) and Parsons (1979)

HSGE: HS from GEOS-3

HSWR: HS from waverider

DISCUSSION

In the report by Fedor et al (1979), an intercomparison of several algorithms is carried out, and GEOS-3 estimates of HS using the different algorithms are in addition compared to buoy measurements. For the algorithm used here, developed by Hayne (1977), the absolute mean difference was calculated to be 0.88 m - based on 27 GEOS-3 passes. Parsons, (1979), also presents an intercomparison of buoy and GEOS-3 measurements using the same algorithm as in our case - see table 2. We find an overall absolute mean difference for Halten, Utsira and Brent to be 1.17 m based on a total of 45 passes. This is higher than the value reported by Fedor et al, and the difference may partly be due to larger space lag between the satellite ground track and the buoy positions in our case. Fedor et al give a correlation coefficient C of 0.667 between buoy and altimeter data while our correlation coefficient based on all three stations is estimated to be 0.91. Thus, we find a higher correlation between buoy and GEOS-3 measurements than Fedor et al. However, our results suggest that GEOS tends to overestimate HS compared to the waverider - see table 2.

In Fedor et al (1979) it is also found that 23% of the satellite measurements are within $\pm$ 0.5m of the buoy measurements for the algorithm in question. In our case we found 28.6% for the Utsira data, 38.5% for the Brent data and 5.6% for the Halten data. Of the 45 GEOS passes near the waveriders, 22.2% of the measurements were within $\pm$ 0.5m of the waverider values.

In fig. 6 on the next page we have plotted (HSGE-HSWR) as a function of HSWR. No clear relationship between sea state and (HSGE-HSWR) can be detected from the scatter plot. But the ratio (HSGE-HSWR)/HSWR seems to decrease as HSWR increases showing that the relative error of the altimeter decreases during higher sea states. This was also expected as the radar altimeter was primarily designed to operate during high sea state conditions.

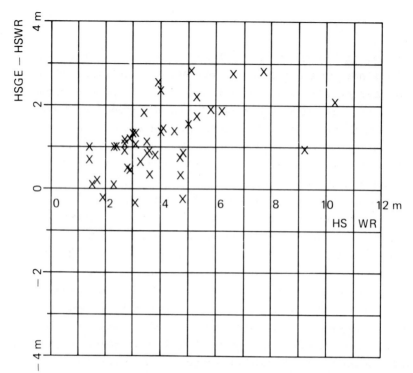

Fig. 6.   The difference in significant wave height from GEOS-3 and the waveriders (HSGE-HSWR) plotted as a function of significant wave height from the waveriders (HSWR).

A natural question to ask now is why the results presented here seem to be different from previous intercomparisons. Though the number of passes is quite limited, the probability that almost all the satellite measurements are greater than the buoy measurements is very small provided both satellite and buoy measurements are sampled from the same probability distribution. The probability that the N first samples are greater than the N last is given by $(N!)^2/(2N!)$. If N = 45 as in our case it is evident that this probability is practically zero. Considering that only 3 of the altimeter measurements were greater than the waverider measurement we conclude that:

1) The GEOS-3 measurements and the waverider measurements are not sampled from the same distribution

or

2) There is an error in the satellite data.

Strictly speaking the waverider should also be questioned, but
we have used data from three buoys and waveriders are regarded
as more reliable than the altimeter.

Excluding the values from Halten in the above intercomparison changes the situation somewhat. We then find the
absolute mean difference to be 0.88 m, the same as found by
Fedor et al, and the correlation coefficient to be 0.81. For
these two stations 33.4% of the data lies within $\pm$ 0.5m of the
buoy data, - the same as found by Fedor et al. (1979) for this
specific algorithm.

Thus, the absolute mean difference and tendency towards
overestimation ceases while only using values from Brent and
Utsira. But the bias in the altimeter data is still present.

One main difference between Brent/Utsira and Halten is the
possible effect of the coastal shadow - see Fig. 4. As
southerly and south westerly winds are common in this area and
noting that most of the GEOS-3 measurements near Halten are
from an area which is further north and further away from the
coast than the buoy measurements, then under suitable wind
direction conditions this increase in fetch might account for
some of the larger discrepancies between GEOS and buoy data at
Halten compared with those at Utsira and Brent. It might also
be the case that the wave field in general is more inhomogeneous in the Halten area than in the Utsira and Brent areas.

In order to check the possible effect of the coastal shadow
in a loose sense, wind directions deduced from weather maps
were used to split the Halten data set into two groups; one in
which the coastal shielding effect was assumed to be of significance.

The absolute mean difference was calculated for both data
sets and it turned out that it was 0.2m higher for the data
assumed to be affected by the shielding effect. However, this
small increase was not regarded as significant for drawing any
firm conclusions. But the increase supports the fact that the
coastal sheltering effect will give rise to higher waves in the
more exposed offshore waters during the appropriate wind
conditions and thus probably will give rise to higher
differences between satellite data and waverider data if the
spatial spread of satellite data is biased in an offshore
direction as in the present study. This is consistent with a
comparison between the waverider at Halten and the ODAS 490

Norwave data buoy situated further off the coast (65°2'N, 7°33'E).

Another way of investigating inhomogeneities in the wave field is to inspect the wave profiles measured by GEOS-3 along the ground track. This was done both for the Halten, Utsira and Brent areas.

It was observed that quite large spatial variations in HS occurred. Differences of around 2 m within 100 or 200 km were not unusual, but it is not possible to say that increasing wave height in a direction away from the coast is a dominating feature. However, the 7 cases at Halten when the wave height increased with distance away from the coast always corresponded to an approaching frontal system or one which had just passed.

This correspondence was not that clearly demonstrated in the Brent and Utsira areas were respectively 5 out of 7, and 2 out of 6 events corresponded to frontal systems.

Regarding the present investigation in relation to those of Parsons (1979), Fedor et al. (1979) and other intercomparisons of in-situ wave/wind measurements and GEOS-3 data, it is worthwhile noting that the latter were carried out using data measured in more open ocean areas whilst our measurements are taken from just off the coast. From table 2 we see that the Brent data gives the least absolute difference, and Brent is also the waverider position furthest off the coast. The satellite also travelled in an east-west direction and as the HS estimates are moving averages, there is a slight possibility that reflections of radar signals from land masses will affect the near coast estimates. More important is probably the change in reflectivity as the ground track passes from land to sea. This change may affect the altimeter so that a bias is introduced in the near coast GEOS-3 data.

As a conclusion we have found some evidence for the presence of spatial non-homogeneities in the wave field - at least in the Halten area - leading to the apparent result that GEOS-3 measurements and the buoy measurements are not sampled from the same probability distribution. In addition some of the bias in the GEOS-3 data may possibly be caused by the effects introduced when the ground track passes from land to sea.

GEOS-3 PASSES OVER METEOROLOGICAL FRONTS

Figs. 7 and 8 show two examples of GEOS-3 passes over meteorological fronts. The surface weather maps presented are based on the 0600 GMT analysis. The satellite passed about two hours later. Thus, the frontal systems would have been positioned further towards north-east than shown on the weather maps. Wind speed observations taken from the weather maps are shown as single points on the plots showing the wind profiles along the surface track.

On February 22nd there may be seen from the weather map to be a series of fronts approaching the Norwegian coast from the south west. The satellite passes over one warm front and proceeds towards another situated to the south of Iceland. Both the wind speed and significant wave height increase as the satellite approaches the first front, straight after which there is a significant drop in both these parameters. The wave height and wind speed then start to grow again as the satellite moves over the second front to the southwest of Iceland.

The next day, February 23rd, the weather situation has worsened and the pass over the frontal system suggests strong activity in the frontal region which gives rise to very strong gradients both in the wind and wave field in a direction away from the coast. However, the measured values of significant wave height and, especially, wind speed on February 23rd, must be considered to be rather high if we take into account that they represent average values over a rather long distance and the observed wind speeds on the weather map. It is very likely that the heavy precipitation and clouds in connection with frontal systems will affect the return radar signal, and this should be taken into consideration. However, no specific algorithm for correcting such errors is to our knowledge yet available, and this is surely a future demand.

These two examples illustrate the ability of satellites to scan the sea surface and provide measured wind and wave profiles which are hardly available by other methods. The present results are promising enough to suppose that satellites may become a powerful tool in the acquisition of real time wind

232

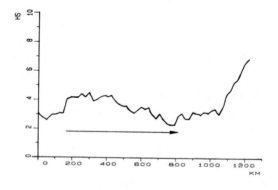

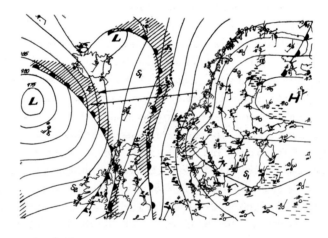

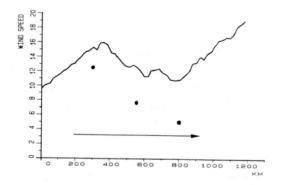

Fig.7. Wind and wave profiles from GEOS-3 on February 22nd 1976 at 0824 GMT. The arrow indicates the satellite propagation direction. Significant wave height (HS) is given in m and wind speed in m/s. The dots represents nearby wind speed observations taken from the surface weather map based on the 0600 GMT analysis.

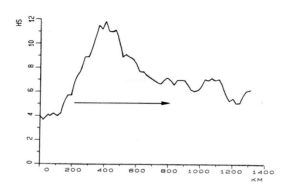

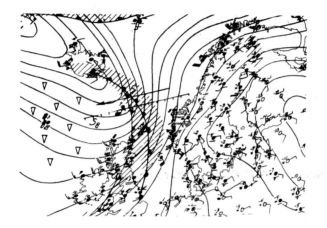

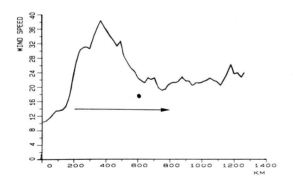

Fig. 8. Wind and wave profiles from GEOS-3 on February 23rd 1976 at 0810 GMT. The arrow indicates the satellite propagation direction. Significant wave height (HS) is given in m and wind speed in m/s. The dot represents a nearby wind speed observation taken from the surface weather map based on the 0600 GMT analysis.

and wave data in the future in particular as measuring techniques and computer algorithms are improved.

ACKNOWLEDGEMENTS

The preparation of this paper has been sponsored by the Royal Norwegian Council for Scientific and Industrial Research (NTNF) through the project "Analysis of Wave Data", 1810.7890 and the joint Norwegian measurement program "Oceanographic Data Acquisition Project" (ODAP). ODAP is financed by Norsk Hydro, Saga Petroleum, Statoil, British Petroleum, Phillips Petroleum and with additional funding from NTNF.

REFERENCES

Barrick, D.E., Determination of Mean Surface Position and Sea State from the Radar Return of a Short-pulse Satellite Altimeter, Sea Surface Topography From Space, Volume 1, NOAA Technical Report ERL 228-AOML, February 1972.

Brown, Gary S., Estimation of Surface Wind Speeds Using Satellite-Borne Radar Measurements at Normal Incidence, J. Geophys. Res. 84 No B8, 3974-3978, 1979.

Fedor, L.S, Godbey, T.W., Gower, J.F.R., Guptill, R., Hayne, G.S., Rufenach, C.L. and Walsh, E.J. Satellite Altimeter Measurements of Sea State - an Algorithm Comparison. J. Geophys. Res., 84, No B8, 3991-4001, 1979.

Hayne, G.S., Initial Development of a Method of Significant Wave height Estimation for GEOS-3, NASA CR-141425, August 1977.

Krogstad, H.E., Gundersen, P., Computer Listings of Wind and Wave data from GEOS-3 North Sea/Norwegian Sea. IKU report P201/2b/81, Continental Shelf Institute, Trondheim.

Krogstad, H.E., Gundersen, P., Wind and Wave Profiles from GEOS 3 North Sea/Norwegian Sea. IKU report P201/2c/81. Continental Shelf Institute, Trondheim.

Parsons C.L., GEOS-3 Wave Height Measurements: An Assessment During High Sea State Conditions in the North Atlantic, J. Geophys. Res. 84, No B8, 4011-420, 1979.

SATELLITE IMAGERY OF BOUNDARY CURRENTS

T. Carstens, T.A. McClimans and J.H. Nilsen
Norwegian Hydrodynamic Laboratories, Trondheim, Norway

ABSTRACT
    Fronts along coastlines are often associated with boundary cur-
rents, and satellite imagery has increased our ability to study
such currents. Their causes are discussed, and it is concluded
that persistent boundary currents are due to the deflection of
runoff by the earth's rotation.
    The much studied Norwegian Coastal Current is used to illu-
strate the dynamics of boundary currents. The basic front distur-
bancies as seen from satellites are reproduced in a small scale
physical model. Although very useful, satellite imagery is frequently limited by clouds, and does not give current speeds.
Moored current meters yield current speeds exceeding present
model predictions, indicating strong meteorological forcing.

REAL AND PSEUDO BOUNDARY CURRENTS

    Fronts are not always associated with longshore currents. Ver-

tical flows are known to produce fronts with significant tempera-

ture jumps, with or without corresponding density jumps. Whether

or not there is a density difference across the front, it is un-

stable and may show configurations not unlike the fronts of the

boundary currents described below. Accordingly, thermal images

can be misinterpreted so that a pseudo boundary current with neg-

ligible longshore flow by an untrained eye is assumed to be a

real boundary current.

    Normally the front of a pseudo current has more random pertur-

bations than the typical eddies and whirls of a real boundary

current. The so-called thermal bar between water masses of equal

density but different temperature is easily deformed, and the

balance of an arrested front between water masses of different

densities is sensitive to stochastic variations.

    If the cause of the front is winddriven upwelling, the stochas-

tic component is readily ascribed to the wind field. If the

driving force is a heat flux, spatial variations are generated for

many reasons, among which topography is an important one.

    In any case the absence of significant shear prevents the

formation of the more predictable disturbances of the real boun-

dary currents.

In very sheltered areas the thermal images may be mapping the temperature field of an undisturbed surface film. This field reveals spatial variance on a small scale and may limit the usefulness of future reductions in pixel size. For the study of coastal currents this source of noise is unimportant, but in small fjords and lakes we may eventually face a problem here.

Boundary currents are readily generated by many different mechanisms and are therefore frequently observed in water bodies of any size. On the geophysical scale we find such currents occurring regularly in lakes, in fjords, in nearshore coastal waters as well as offshore along continental shelves. In fact, boundary currents seem to be the rule rather than the exception. The exceptions are those lakes in which the temperature gradients vanish for a few days each year and those shallow seas in which the density gradients vanish seasonally due to cooling and ice formation. In the discussion below we have excluded the large scale oceanic circulations along continental shelves. Satellite pictures have provided an overwhelming amount of evidence to prove the occurrence of boundary currents.

CAUSES OF BOUNDARY CURRENTS

Runoff

When a river discharges into a lake or on an open coast, it will normally remain an identifiable entity for a considerable distance from the source, which is the river mouth. As soon as the solid boundary (or, in the case of alluvials, not so solid boundary) is replaced by a fluid interface, mixing begins. In cases with weak tides the higher turbulence of the river, compared to the recipient, drives a one-way entrainment process by which the surrounding water is pulled into the now free-floating river. Gradually the composition of the river water changes so that it resembles more and more the receiving water. In many cases this process begins already far inland when a salt wedge lifts the river away from its bed.

In cases with strong tides there is a two-way horizontal exchange by turbulent diffusion. Although the salinity at the mouth may approach that of the sea, there will normally be enough of a density deficit to make detection of the river possible even far away from the mouth.

Jet deflection

For a river to form a boundary current in a lake or along a coast, it will have to change its course. Changes of direction occur for a number of reasons, among which cross flow is perhaps the most obvious one. However, we shall point out the other reasons as well and demonstrate that in the final analysis the earth's rotation provides a satisfactory explanation to our observations.

The momentum balance for the laterally constrained river flow is usually written in the axial direction only. The lateral forces are small in comparison and their consequences negligible with a few exceptions. One such exception is a bend in an alluvial river, where centrifugal forces acquire importance. In any case the balance is easily achieved by a lateral surface slope compensating for the possible imbalance of the other lateral forces: Coriolis force, centrifugal force, wind stress and lateral friction.

When the river leaves its mouth, the lateral slope can no longer be maintained. The river is thrown into a lateral imbalance which causes it to deflect and to spread side-ways.

River trajectories have been computed for many different cases. The simplest case is that of a discharge for a stagnant sea as shown in Fig. 1.

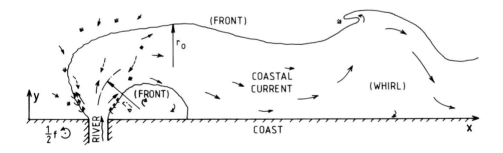

Fig. 1. Coriolis deflection of a river plume to form a boundary current. (McClimans, 1983)

After a travel of about

$$r = u/f \tag{1}$$

the river has turned 90° and become parallel to the coast. Here u is the speed of the jet and f is the so-called Coriolis para-

meter. Depending on the width of the river, it will continue to
bend until it makes contact with the coast. For large river
widths B >> r the outflow will be longshore within a distance
x = B. For B << r the flow makes contact at x = 2r. In the
northern hemisphere this reattachment occurs to the right of the
mouth, while in the southern hemisphere it would be to the left.

Downstream of the point of reattachment geostrophic equilibrium
is established by means of a surface slope in the offshore direction. The equilibrium width of this subsequent coastal (boundary)
current is about two times the Rossby deformation radius r_i given
by

$$r_i = \frac{c_i}{f} \tag{2}$$

$$c_i = (g' \frac{h_1 h_2}{h_1 + h_2})^{1/2} \tag{3}$$

where c_i is the phase speed of long internal waves, h_1 and h_2 the
depth of the upper and lower layer, respectively and g' is the reduced gravity

$$g' = \frac{\Delta\rho}{\rho} g \tag{4}$$

Here $\Delta\rho$ is the density difference between the current density ρ
and the density of the ocean.

Geometrical wall effects

There are several deflecting forces other than the Coriolis
force. A common case is the wall proximity: When the angle between the jet and the shore differs from 90°, the entrainment becomes asymmetric. As a result a lateral pressure difference develops, and the jet is pulled towards the nearest boundary. If
the angle α is small, 20 or 30° say, the jet will reattach to the
shore. This effect is known as the Coanda effect in fluidics.

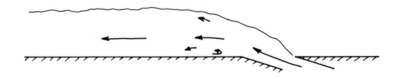

Fig. 2. The wall proximity may cause a boundary jet.

It may be argued that constraints on the replenishment of entrained water would produce a lowering of the water surface at
the wall, causing the jet to hug the shore. Another geometric
constraint akin to a close shore is a sloping bottom. This may
cause a bottom boundary current which evades remote sensors.

Cross flow

The case of a jet in a cross flow has been dealt with extensi-
vely (see e.g. Fan, 1967).

Most researchers have related the drag force on the jet to
that on an equivalent immersed solid body. Offhand, the definition of a jet surface on which to apply the pressure and shear
distribution of such a drag force would seem obscure. However,
the experimental evidence is quite conclusive. The jet behaves
much like a floating (or submerged) solid body. The observed
flow pattern reveals that the jet changes not only its course,
but also its shape. In some cases a circular jet is transformed
into a kidney-shaped jet as seen in Fig. 3, and eventuelly two
jets may be formed as in the case of a buoyant submerged plume
in a cross-flow.

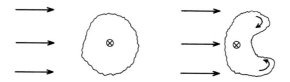

Fig. 3. Transformation of a circular jet in cross-flow.

Wind

The river jets of the preceding examples are all found in the
upper few metres of the sea. They are therefore heavily influenced by wind drag directly on their exposed free surface. But
they are also indirectly affected by the wind-induced surface
current in the sea on which the river floats.

Local wind. The way the wind pushes a nearshore river plume
around is easily observed anywhere. Onshore winds shift the
boundary or front towards the shore and produce very distinct
fronts with sharp horizontal density gradients. Offshore winds
force the lighter water of the boundary current to spread thin
on top of the heavier sea water. The boundary becomes blurred

and irregular.

Seasonal wind. The local wind effects described above are repeated on a larger scale by the monsoon character of most coastal climates. Seasonal reversals of the prevailing winds are common, and so are seasonal changes in boundary currents.

The thickening and thinning of a boundary current are closely associated with the seasonal up- and down-wellings. From a monitoring point of view this is important as other tracers than water temperature become available, e.g. chlorophyll concentration.

Pulsed wind excitation. A common occurrence of potentially great importance to boundary currents is the pulsating or valve action of storms. Winds from a certain sector may stop or decelerate the current, more or less as the turning of a valve in a pipe. Upstream, the water piles up until the wind ceases, which is analogous to reopening the valve. The detained water now rushes out, forming a pulse in the boundary current.

While the qualitative aspects of this kind of wind forcing are fairly evident particularly for storms of a few days duration, there is a lack of quantitative data. Since extreme values of flux and velocity in a boundary current are expected to be associated with such outbreaks, a closer look at wind effects should yield important information.

In case the current transports pollutants, a detention by wind blockage as described above may cause accumulation upstream (down wind). Concentration peaks are thus expected to be correlated with wind. For decaying pollutants such as waste heat the correlation with the wind will be different.

Direction of deflection

All deflecting mechanisms other than the Coriolis force work equally well in both horizontal directions. Topographical effects are essentially random and can turn the jet left or right. Naturally both crossflow and wind forcing have a predominating direction determined by the local climate. But this direction may again be either to the right or to the left. As a conclusion we can state that the most persistent boundary currents are generated by the rotation of the earth.

THE NORWEGIAN COASTAL CURRENT (NCC)

The specific example of boundary currents to be treated here
deals with the outflow from two semi-enclosed seas: The Baltic
and the North Sea.   Contrary to the Mediterranean these two seas
have more runoff than evaporation, so there is a net buoyancy
flux out.   This flux leaves in the form of a geostrophic surface
flow along the Norwegian coast.

Helland-Hansen & Nansen's 80 year old map contains all the conventional knowledge of average conditions.   This has been updated
by Braaten & Sætre (1973) as shown in Fig. 4.   Helland-Hansen &
Nansen also had some appreciation of the variability and stated
that the coastal current is wider and thinner in the summer when
the buoyancy flux is high.   This is readily derived from the geostrophic balance of the Coriolis force with the lateral pressure
gradient.   Margules' formula for depth H and width B of a wedge-
shaped buoyant coastal current is (see Fig. 5)

$$\frac{H}{B} = \frac{fu}{g'} \tag{5}$$

This formula appears to describe the gross features of the NCC
quite well.   Much of the earlier interest in the NCC has been due
to its effects on sill fjords.   In the winter the thickness of the
NCC may exceed the sill depth of many fjords, blocking the flow
of atlantic water to the deeper parts of the fjord basins.   In
the spring/summer the reduced density/increased buoyancy will
cause a reduction in the thickness of the NCC and allow atlantic
water to enter and replace polluted fjord water.   Thus we may
have an annual clean up that resets the stage for the next period
of blocking.

Observations of puzzling episodes like flow reversals have
been reported from some of the many cruises off the west coast of
Norway.   They have been proposed to be rings like those shed by
the Gulfstream.   However, there was very patchy evidence of this
before the advent of satellite remote sensing.

Recent observations

A renewed interest in the NCC has addressed issues of vital
importance to the nation.   The first impetus came from the pollution threat.   Very conspicuous are plastics that are almost in-
destructible and chemicals like PVC, which have been observed at

242

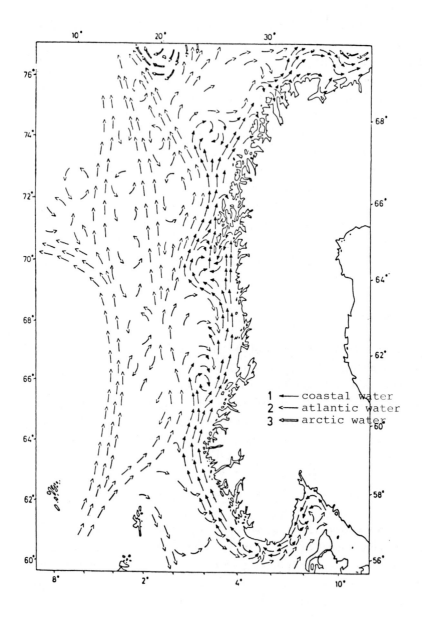

Fig. 4.   Residual currents off the coast of Norway.

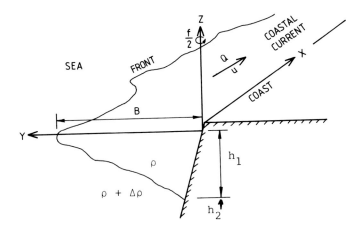

Fig. 5.   Schematic of a baroclinic coastal current.

Spitzbergen and in the Barents Sea, several thousand km from the
nearest source.

   Our coast is threatened not only by land-based pollution, but
an increasing threat of oil spills from offshore oil fields is
becoming more apparent. Any oil drift model needs the advection
fields. Perhaps the NCC would act as a boom and collect oil at
its flank or front, where there would be a convergence?

   Another concern has been the possibility that man-made changes
in the NCC would affect the fish population. Several of our eco-
nomically most important fisheries, and in particular the cod,
use the NCC for a convenient free ride from the spawning ground
to the Barents Sea where the juveniles grow up. Furthermore the
supply of food for the fries comes by the NCC. Some biologists
argue that any change in flux or composition of the waters that
make up the NCC may have an adverse effect on the fisheries. On
the other hand, the natural variability in the NCC is formidable,
and any man-made contribution to the variance is likely to be in-
significant. The mean annual runoff from Norway is about
10^4 m^3/s. It is typically diluted 10 - 12 times to a buoyancy of
about 2 kg/m^3, i.e. the dilution parameter, defined as the ratio

of total flux Q to runoff Q_f

$$P = \frac{Q}{Q_f} = \frac{s_o}{s_o - s} \tag{6}$$

With $s_o = 35\ °/_{oo}$ and $s = 32\ °/_{oo}$ eqn. (6) gives P = 12.

So the contribution of the local runoff to the transport in the NCC is about $10^5\ m^3/s$ or 0.1 SV. This tallies well with observations since most of the fresh water in the NCC comes from the rest of northern Europe. An average of about 0.4 SV gives an annual outflow in the NCC of 12 000 km^3. In comparison the annual exchange flows in the Mediterranean were 53 000 km^3 in and 50 500 km^3 out, with a net inflow of 2 500 km^3 (Ejenidi, personal communication).

The ensuing discussions triggered two significant activities. First, a four year effort by all our oceanographers who studied the NCC intensively from 77 to 81 and updated our knowledge consierably and second, the commissioning of an experimental facility at the River and Harbour Laboratory in Trondheim; a 5 m diameter rotating basin.

The NCC project uncovered more variability than we had previously been aware of. Cases of meteorological forcing were evident. Wind from the west was piling up water in the Skagerrak, and when the wind stopped there was a surge of released water flowing around the southern tip of Norway. Although this phenomenon was first documented by Eggvin (1940) and has certainly been known by local fishermen for centuries, it was considered a rare event.

But perhaps the most exciting was the week back in 1978 when we started the rotating model and obtained the strangest behaviour of a boundary current flowing along a straight wall. We saw many kinds of waves and disturbances along the free boundary of the current. Fig. 6 shows the type of flow patterns simulated in the rotating basin. If we released a fast jet with initial densimetric Froude number

$$Fr' = \frac{u}{c_i} \tag{7}$$

we got backward breaking waves much as in a non-rotating case of overcurrents or undercurrents. When the initial flow was sub-

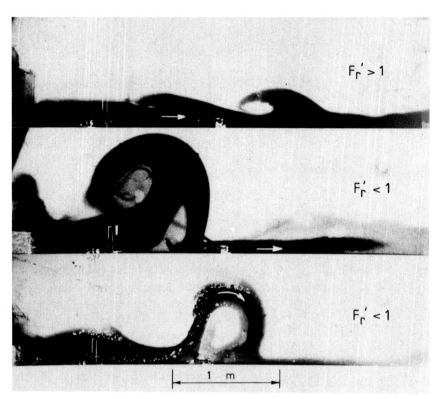

$F_r' > 1$

$F_r' < 1$

$F_r' < 1$

1 m

Fig. 6. Laboratory simulation of baroclinic coastal currents. (Vinger & McClimans, 1980).

critical, Fr' < 1, the boundary current separated from the wall, made a loop and reattached further downstream. We looked in amazement: Was this real? Were we modeling nature or observing a model effect?

A few days later we received the first satellite thermal image for comparison. It was a very crude image with poor resolution. But there they were, the bulges that we had just produced in the lab! Since then we have been snowed down with pictures from a number of satellites processed by a number of methods. A receiving station was established in Tromsø a few years ago, and we subscribe to quick look pictures. For clear weather conditons we order a gray tone enhanced thermal image as exemplified in Fig. 7. If we find something interesting we order a computer compatible digital tape and process it on the ICAN screen in Trondheim to

Fig. 7. Gray tone enhanced thermal image of the Norwegian coastal
Current. NOOA-6 13-4-81.

give in false colours a clear representation of the flow. An example from October 1982 is shown in black and white in Fig. 8a.
The similarity with the laboratory flows for Fr' < 1 is apparent.
The flow develops into a cyclonic eddy to the left, and a whirl
to the right as seen in Fig. 8a. This nomenclature has been used
to distinguish between two fundamentally different flows: The

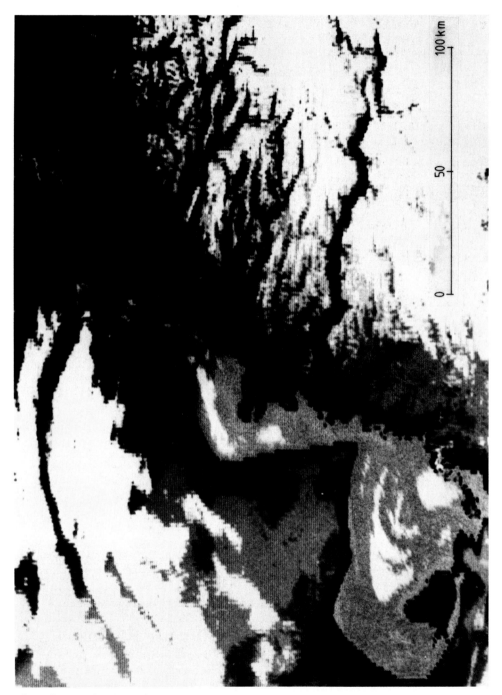

Fig. 8a. Black and white copy of colour enchanced thermal image
of the Norwegian coastal Current along the west coast.

248

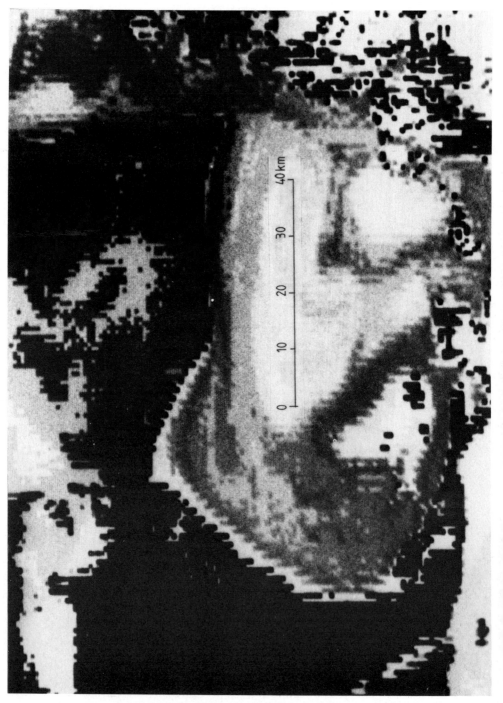

Fig. 8b.   Close-up of the eddy and whirl in fig. 8a.

eddy promotes mixing with the external sea while in the whirl, the coastal jet turns into itself causing a large pool or ring of coastal water. In Fig. 8b we are zooming in on the most interesting details.

For the laboratory simulation of such large scale flows the basic laws that should be satisfied are

$$Fr_r = Fr'_r = Ro_r = 1 \tag{8}$$

Here the subscript r refers to the ratio between natural and laboratory numbers. Fr is the Froude number, Fr' is the densimetric Froude number and

$$Ro = u/fB \tag{9}$$

is the Rossby number. This can be done without problems. Friction is the notorious headache in laboratory simulations, where the ratio of drag coefficients $Cd_r > 1$. However, in a distorted model $((B/H)_r < 1)$ the importance of interfacial friction is reduced nearly to the conditions in nature.

Advection of oil spills

We went to the Department of Environment with our discoveries and obtained a small contract to find what whirls might do to oil drift. The idea was to establish a forecasting routine that would improve existing routines and which could be activated in case of a spill.

An example of an oil spill being blown across a coastal current whirl is shown in Fig. 9 (McClimans & Nilsen, 1983) and in Table 1. An average whirl could speed up the landfall of a spill by 1 day from 3 to 2 and reduce the drift to 15 km. On the other hand it could also delay the land-fall by 2 days and deliver the oil 90 km away. The authorities apparently have not yet found these and other possibilities for improved forecasts sufficiently interesting. Pollution, however, was only one of several issues concerned with the NCC.

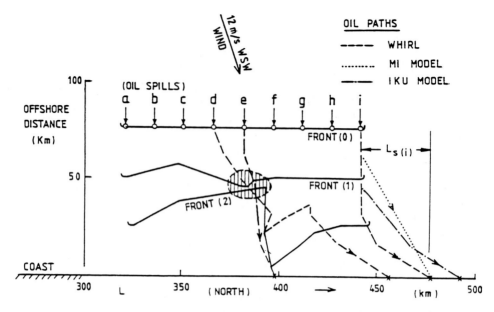

Fig. 9. Position of an oil slick originally (day 0) at positions a-i illustrated as a front after 1 and 2 days. The current field is that of an average whirl originally (day 0) centered at L = 350 km and moving 15 km/day northward. Particle paths for three oil patches (at d, e and i) are drawn with dashed lines. A convergence zone for the oil is shaded. (Mc Climans & Nilsen, 1982)

TABLE 1

release	ΔT days	ΔL km
a	3	37
b	3.5	49
c	5.2	78
d	4.8	90
e	2.1	15
f	2.5	22
g	2.8	22
h	2.9	32
i	3	34
mean	3.3	42
drift model 1	3	34
drift model 2	3	50

Design and operation of offshore structures

In March 1981 the NCC itself came to our rescue by hitting a moored current meter with unprecedented high velocities (Fig. 10). The current meter rig was part of a monitoring programme at the

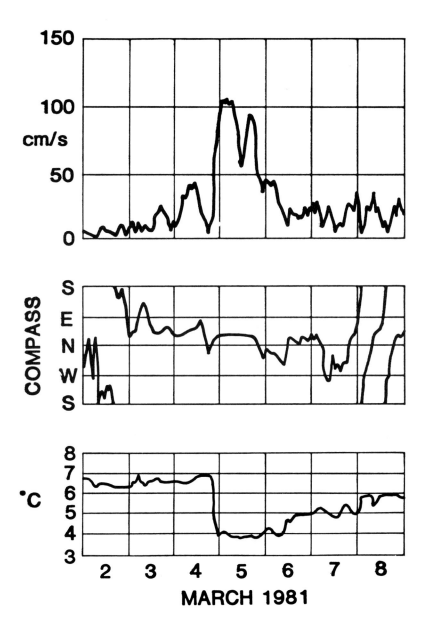

Fig. 10. Current velocity and temperature measurements at 2 km depth ca 20 km offshore (Courtesy A/S Norske Shell)

Troll Field, believed to be the largest offshore gas field in the North Sea.

The time history observed is neatly explained. If the speed increase is directed towards NE it should be followed by a period of strong constant N current before it veers to NW and decelerates. This indicates that the moored current meter is near the outer edge of a whirl. At the same time, a drop in temperature is registered as the colder coastal water passes the moored current meter.

Routine statistics, f.inst. exceedance or duration diagrams form the basis for decision-making. Risk levels are assumed and this leads to probability of design events and eventually to the selection of the design event itself. In the case of the offshore bursts of coastal water, it is felt that satellite images can help fill in the statistics.

The laboratory simulations revealed a relationship between the maximum current speed and the phase speed of a northward propagating whirl. Use of consecutive thermal images has been shown to give a reasonable phase speed compared to the laboratory results (Johannessen & Mork, 1979). Thus it is possible to use satellite thermal images for several quantitative analyses.

It should be remarked that the laboratory simulations did not include wind or other variability. Our predictions were that the disturbances would extend 100 km offshore and cause velocities of 1 m/s or about twice what was expected in the absence of wind. After a rather short observation period of 2 years much larger currents have been observed.

OTHER COASTAL CURRENTS

Our example with the Baltic outflow turning right and becoming the Norwegian Coastal Current is but one of a widely distributed type of runoff-generated boundary currents. Fig. 11 shows a map of the winter currents in the East China Sea. The runoff to Pohai Bay starts the boundary current which is joined by other rivers and runs all along the Chinese coast.

Fig. 12 shows a satellite picture of the Sea of Japan. As described by Sugimura et al. in the present volume, the Tsushima Current can be seen to flow northward along the east coast of Japan as a boundary current. At the same time pseudo boundary currents are visible along the Sibirian coast.

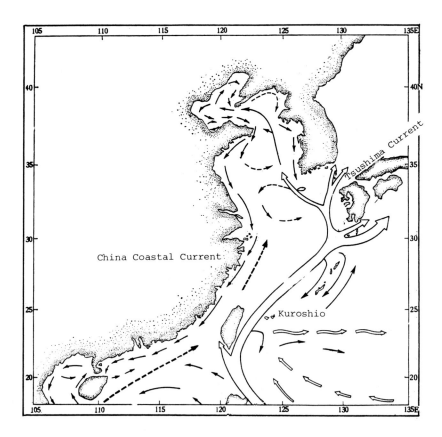

Fig. 11.   Winter currents in the East China Sea, with boundary
current flowing southward. After Guan and Mao (1982).

BOUNDARY CURRENTS IN FJORDS

Rivers discharging to large fjords also reveal boundary currents held shoreward by the influence of the earth's rotation.
Fig. 13 shows a black and white copy of a colour enhanced
LANDSAT picture revealing the pattern of outflow of glacial
flour in the surface waters of Gaupnefjord in the inner reaches of
the Sognefjord. The resolution of this picture is 80 m. As the
resolution improves, the application of remote sensing increases.

Fig. 12. Gray tone enhanced thermal image of the Sea of Japan, with boundary current flowing northward along the east coast of Japan. (Courtesy Remote Sensing Technology Center of Japan).

CONCLUSIONS

The use of remote sensing can provide quantitative measures for boundary current processes in lakes, rivers, fjords and the oceans. Both termal images and optical windows provide useful data for clear-weather situations. As the resolution improves (smaller pixels) more classes of boundary processes will be subjected to remote sensing and analysis.

ACKNOWLEDGEMENTS

This work has been supported by the fund of License Fees.

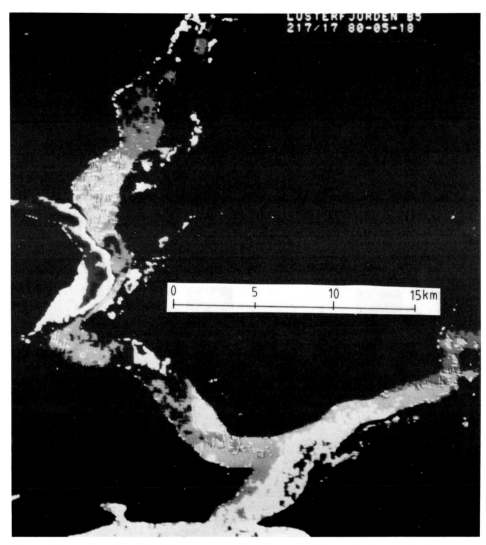

Fig. 13. Black and white copy of colour enhanced thermal image of a narrow fjord.

REFERENCES

Braaten, B.R. and Sætre, R., 1973. Farming salmon in Norwegian coastal waters. Fisken i havet, ser. B (9) (In Norwegian)

Eggvin, J., 1940. The movements of a cold water front. Rep. Norw. Fishery Mar. Invest. 7:1-151.

Fan, L.N., 1967. Turbulent buoyant jets into stratified or flowing ambient fluids. Cal. Inst. of Tech., Report N° KH-R-15.

Guan Bingxian and Mao Hanli, 1982. A note on circulation of the East China Sea. C.J. of Oceanology and Limnology, Vol. 1, N° 1.

Helland-Hansen, B. and Nansen, F., 1909. The Norwegian Sea. Report on Norwegian Fishery and Marine Investigations, Vol. 2.

Johannessen, O.M. and Mork, M., 1979. Remote sensing experiment in the Norwegian coastal waters. Spring 1979. Samarbeidsprosjektet den Norske Kyststrøm. Report 3/79. Geophysical Institute, University of Bergen.

McClimans, T.A., 1983. Laboratory simulation of river plumes and coastal currents. ASME symposium on modeling of environmental flow (in press).

McClimans, T.A. and Nilsen, J.H., 1982. Whirls in the Norwegian Coastal Current and their importance for evaluating oil pollution. Symposium on physical processes related to oil movements in the marine environment. Tväarminne, Finland. November, 1982.

Vinger, Å. and McClimans, T.A., 1980. Laboratory studies of baroclinic coastal currents along a straight, vertical coastline. River and Harbour Laboratory. Report STF60 A80081.

TURBULENCE DISTRIBUTION OFF USHANT ISLAND MEASURED BY THE OSUREM HF RADAR

P. PIAU and C. BLANCHET[1]

[1]Institut Français du Pétrole, 1 à 4, avenue de Bois Préau,
92506 Rueil-Malmaison Cedex, France.

ABSTRACT

The HF radar OSUREM was installed for the spring of 1982 on the coasts of Brittany (France) in order to measure wind and waves off Ushant Island. This type of radar provides electromagnetic spectra. The part of these spectra containing the most energy is the so called first-order spectrum composed of two lines which are generated by a Bragg reflection process. The width of these lines is representative of the energy distribution of turbulent currents in the resolution cell, typically ten kilometers wide. In the experiment described here, the width of the first-order line was suprisingly great, due to the strong tide currents encountered in this area. We can extract the current distribution in the area from the shape of the lines, depending on the time and strength of the tide.

INTRODUCTION

The use of high frequency (HF) electromagnetic waves for the remote sensing of the sea surface has been increasing in recent years. HF radars measure the Doppler spectra of the return power coming back from the sea. If we use a narrow-beam HF radar, the main feature in the Doppler spectrum is the so-called first-order spectrum. It is composed of two lines the mean frequency of which is related to the sea current.

In a lot of geographic areas, it is very easy to extract this current from the first-order Doppler spectrum. Off Ushant Island, we will show that this first-order spectrum is more complicated, due to the turbulent current distribution in the resolution cell of the radar. We will not only extract the current speed with good accuracy but also obtain a measurement of the variance of the current speed in the direction of the radar beam.

1. DESCRIPTION OF THE OSUREM HF RADAR

1.1 What is a narrow-beam ground-wave radar ?

OSUREM is a ground-wave radar. This means that the electromagnetic waves follow the sea surface in going to the measurement area and in coming back to the radar. Another way of propagation is reflection by the ionosphere which is used by sky-wave radars.

Sky-wave radars may measure at distances of the order of some thousand kilometers. But for the moment they cannot obtain spectra good enough to provide accurate wave and current measurements.

Ground-wave radar provides very good Doppler spectra with good time coverage. It does not depend directly on the variations of the characteristics of the ionosphere.

OSUREM is also a narrow-beam radar. The resolution cell is a small one (typically 15 km x 15 km) and is situated far away from the radar (from 20 to 150 km). The azimuthal resolution is obtained by the narrow-beam of the reception antenna. This antenna is 180 meters long and composed of 16 little antennas spaced 12 meters apart.

1.2   Principle of the measurement

The OSUREM radar provides Doppler spectra like the one in Figure 3. This spectrum uses a dB vertical scale. It follows that the three main peaks in the spectrum contain a lot more energy than all of the remaining power. The central peak at the Doppler zero is an electronic artefact. The two peaks with quasi opposite Doppler frequency $\pm f_B$ ($\pm.23$ Hz) compose the first-order spectrum. They are called the Bragg lines. The remaining power in the spectrum is called the second-order spectrum.

1.2.1   The first-order spectrum

Its name comes from an analogy with the Bragg diffraction in crystals. The electromagnetic incident wave (Fig. 1) with wavenumber $\vec{k}_i$ is backscattered by the water waves whose wavenumber is:

$$\varepsilon \, \vec{k}_B = - \, \varepsilon \, 2 \, \vec{k}_i, \qquad \varepsilon = \pm 1$$

Those water waves have the same direction as the radar beam and a wavelength half of the electromagnetic wavelength.

The Bragg waves are gravity water waves, and their velocity is:

$$V = \sqrt{\frac{g}{k_B}}$$

The Bragg Doppler frequency is the Bragg water wave frequency

$$f_B = \frac{\sqrt{g k_B}}{2\pi}$$

Expressed with radar characteristics, this frequency becomes:

$$f_B = \sqrt{\frac{g \, F_r}{\pi c}}$$

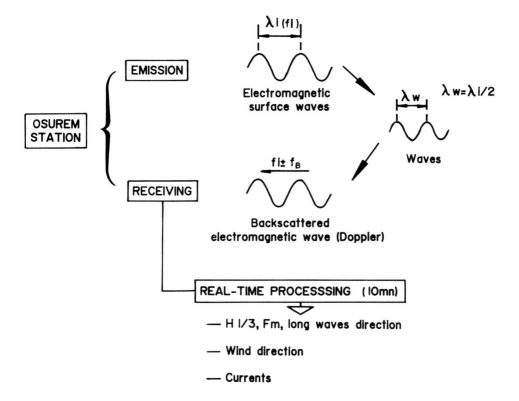

Fig. 1. Principle of HF radar

with F_r = radar frequency, 3 to 30 MHz ; c = speed of the light.

The power of a Bragg line is proportional to the value of the directional sea spectrum $S(\vec{k})$ at the point $\varepsilon \, \vec{k}_B$, with ε being the sign of the Doppler frequency. ε is positive for the waves traveling toward the radar. Then the first-order Doppler spectrum provides the velocity and power of two kinds of waves with wavenumbers $+ \vec{k}_B$ and $- \vec{k}_B$. So it is used to calculate the wind direction and the current speed. The wind direction is inferred from the direction of Bragg waves which are high-frequency waves strongly dependent on the wind. This direction is obtained from the ratio $S(\vec{k}_B)/S(-\vec{k}_B)$ which is maximum in upwind conditions (20 to 30 dB) and equal to 0 dB in crosswind conditions.

This method was used with groundwave radar [1] and with skywave radar [2]. There is an ambiguity because two symmetrical directions in relation to the radar beam give the same wind direction. The accuracy of the method is improved and the ambiguity suppressed by the use of three radar beams [3].

The measurement of the current characteristics is deduced from the position in frequency of the Bragg lines, which is not always $\pm \, f_B$. This point will be developed later.

1.2.2 The second-order spectrum

This spectrum is the result of second-order interactions between Fourier components of the height of the sea surface. These interactions appear in the equations for hydrodynamics and electromagnetism. It follows that two waves are involved in one elementary interaction. For the hydrodynamic part, the radar measures the second order waterwave spectrum with wavenumber $\pm \vec{k}_B$. This spectrum contains power for every Doppler frequency. So it differs from the first-order water wave spectrum which is a Dirac function for a given wavenumber $\vec{k}_B$. The electromagnetic part also gives a continuous spectrum, which is added to the hydrodynamic one.

The shape of the second-order spectrum, starting from a Bragg line, is directly related to the shape of the low-frequency part of the waterwave spectrum. For one peak in the waterwave spectrum, there are four peaks in the radar spectrum, on each side of each Bragg line.

A lot of work has been done to extract the waterwave directional spectrum from the radar Doppler spectrum.

Narrow-beam radar is the only one which can measure at a great distance from the radar. But for this type of radar, the maximum power in the Doppler spectrum is not always provided by the dominant direction of the low frequency waves. This is the case for waves traveling perpendicular to the radar beam. This problem has been solved for the OSUREM radar. This radar provides:

- the direction of the high frequency waterwave
- the height of the high frequency waterwave
- the direction of the low-frequency waterwave with an accuracy of $10°$
- the spectrum of the low-frequency waterwave. The accuracy is the frequency of the maximum which is the same as with a buoy
- the total height of the waterwave with an accuracy of 20% in all conditions.

Extensive experiments were performed with H 1/3 from 2 to 7 meters.

Figure 2 shows a wave spectrum obtained by the OSUREM radar and a Datawell buoy in a complicated sea, with the wind being perpendicular to the low-frequency waterwaves.

2.   TURBULENCE MEASUREMENT : PRINCIPLE

The current characteristics are deduced from the first-order Bragg lines. Usually the current speed is related to the position of the Bragg lines. This paper shows that the shape of the Bragg lines gives the spatial distribution of the turbulent current.

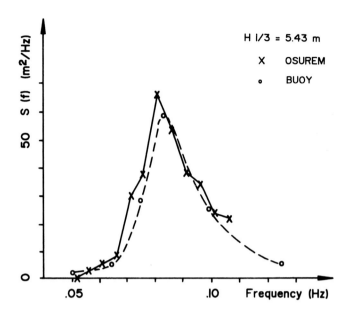

Fig. 2. Comparison of long wave spectra obtained with OSUREM radar and a Datawell buoy.

2.1 Mean current measurement

If the mean current in the resolution cell of the radar has a non zero component in the direction of the radar beam, the entire electromagnetic spectrum is shifted by:

$$\Delta f_{cr} = -2 \frac{V_{cr} F_r}{c}$$

Δf_{cr} : Doppler shift due to the mean current.
V_{cr} : speed of the mean current.
c/F_r : radar wavelength.

This type of simple measurement was the first application of HR radar, implemented in the U.S.A. [5] and in France [6].

Since the Bragg lines have a non zero width, it is necessary to define their position with the greatest accuracy. The centroids of the Bragg lines bounded by the values 3 dB under the maximum, may be used [7]. In this case, the standard deviation of the frequency estimation is given by:

$$\sigma_f = \sqrt{N} \frac{1}{2T}$$

T : total duration of the measurement

$$N = \frac{\Delta f}{1/T}$$

Δf : width of the peak

The accuracy is independent of the way of averaging incoherent spectra and depends only on the total duration T. The accuracy varies with the width of the peak.

Usually, the Bragg lines are very sharp. All the energy is concentraded in a few samples.Figure 3 shows one of this kind of spectra obtained during the Marsen experiment in the North Sea with OSUREM radar. Spectra like this one were also obtained off the Shetland Islands, in the Bay of Biscay or in the Mediterranean Sea.

But the spectrum shown in Figure 4 was measured off Ushant Island. The energy distribution in the Bragg line is wide and cannot be assumed to be Gaussian. So we choose for this experiment the limits of the Bragg lines at 10 dB under the maximum. This limit is chosen in relation to the value of the first side lobe of the reception antenna, which is - 15 dB in relation to the main beam.

We will see later that we may extract an accurate tide stream from the Doppler spectra with this method.

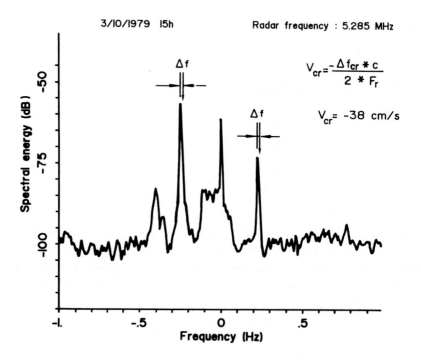

Fig. 3. Principle of the mean current measurement with a narrow beam HF radar.

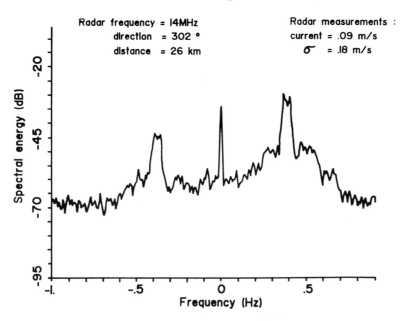

5/5/1982 19h45 (3h after high water at Brest)

Radar frequency = 14MHz
direction = 302 °
distance = 26 km

Radar measurements :
current = .09 m/s
σ = .18 m/s

Fig. 4. Typical Doppler spectrum obtained with OSUREM radar off Ushant Island. The Bragg lines are very wide.

There are two measurements of the current, with the positive and the negative Bragg lines. We use only spectra with a difference between the two measurements lower than the theoretical standard deviation defined earlier. This is the case in general. It is an important result obtained with HF radars : the Bragg lines are always nearly at the position deduced from the dispersion equation for the waterwaves, even with very rough seas. The current is taken as the mean of the two measurements.

The depth at which the current is measured is a fraction of the radar wavelength. At 10 MHz, this depth is a few meters.

2.2   Turbulence measurement

As we have seen, the Bragg lines are far from a Dirac function for the Ushant experiment. Some work was done on using the width of the Bragg lines to extract wind speed. It seems that this width is essentially due to the advection of the Bragg waves by inhomogeneous currents in the resolution cell of the radar. These currents may be orbital motions of long waves or turbulent surface currents. For the Ushant experiment, orbital motions are one order of magnitude less than the Doppler width of the Bragg lines. So we expect this width to be related to the turbulent surface current.

The power density at a given frequency in a Bragg line is representative of the sea area occupied by the current with the corresponding speed. So what is measured is the sea area where the current has this given speed. More exactly it is the component of the turbulent current along the radar beam which is used. The power received from a sea area (Fig. 5) is proportional to the Bragg water-wave energy in the area and to the area. So the total power in the spectrum for a given Doppler shift is proportional to the total area where the current speed corresponds to this Doppler shift. But this is true only if there is no correlation between wave power and current speed. This point will be discussed in the last section of the paper.

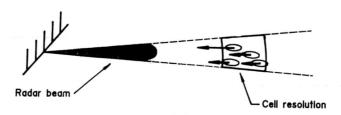

Fig. 5. Principle of spatial turbulence measurement.

From the standard deviation of the Bragg lines considered as a distribution, we deduce a characteristic of the turbulent current, i.e. the standard deviation of the spatial distribution of its component in the radar beam direction. As is well known in turbulence theory, this spatial measurement is very different from conventional point measurement evolving with time. An attempt was made to study the effect of the measurement duration on the results. There was no significant difference between 3 and 20 minutes.

3. THE USHANT EXPERIMENT

3.1 The radar site

The OSUREM radar was installed for three months on the Brittany coast opposite Ushant Island (Fig. 6). The radar was looking in five directions. The 302° direction met a point called point 429. At that point, the tide stream had been measured by the French Hydrographic Office (SHOM).

3.2 The tide off Ushant Island

The Kelvin wave coming from the south enters the channel as a progressive wave (Fig. 7). The tide stream may be represented as a horizontal ellipse. The current turns clockwise with a semidiurnal period (Fig. 8). The closer we get to the coast, the flatter the ellipse becomes. Far from the coast, at about

100 km, the ellipse is almost round.

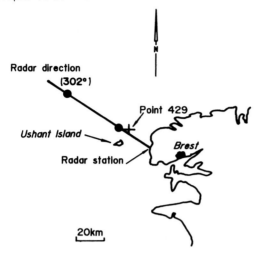

Fig. 6. The measurement area off Ushant Island.

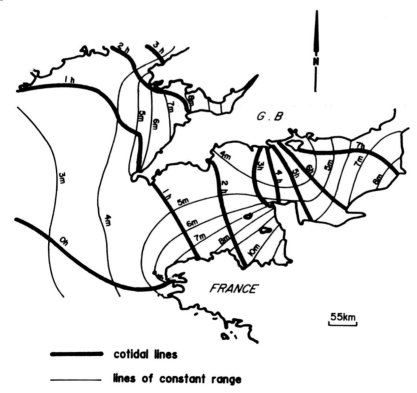

Fig. 7. The tide off the coast of Brittany.

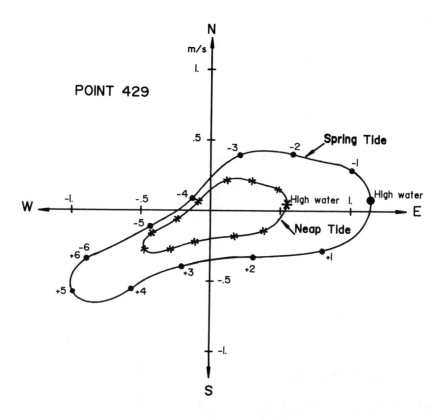

Fig. 8. Tidal ellipses at point 429 (from the tide tables of SHOM).

SHOM provides two ellipses measured for neap and spring tides. These ellipses
are measured in a short campaign, so the accuracy is not very good, and the
direction of the tide stream is assumed to be independent of the tide power.
In France, tidal power is scaled with a coefficient which is proportional to
the tide generating force. The mean neap tide has the coefficient 45, and the
mean spring tide the coefficient 95, independently of the absolute value of the
current speed or of the tide height. Off Ushant Island, the current speed is
greater than 1 m/s for the mean spring tide and in an area 50 km wide. All this
area is included in the continental shelf around 100 meters deep. During a
semidiurnal period, a water particle follows an ellipse 10 km wide.
    So in the area the tide is very strong and is turning around an island.
The depth is great enough not to dissipate turbulence, but shallow enough to
generate turbulence with its main topography.

3.3   Meteorological and oceanographical features
    The dominant wind comes from the west, from the Atlantic Ocean. The strong-

est winds usually turn from southwest to northwest as the depression moves northward. During the Ushant experiment in the spring of 1982, the winds were almost gentle in all directions.

The thermocline is established in July. At that time there are a lot of internal waves coming from the southwest, and oceanic fronts appear in the area. But in the spring, we may assume that the thermocline is not established, even if we have not measured it.

4. MEASUREMENT OF THE TIDAL STREAM

As said earlier, the mean current is deduced from the barycenters of the Bragg lines. The measurement is accepted if the two Bragg lines give the same current speed. The results are compared to the tide table published by SHOM [8]. There are several points of measurement of the tidal stream. Point 429 is close to a radar cell (Fig. 6). The depth of the SHOM measurement is 3 to 5 meters.

To compare tide streams independently of the tide ranges, we normalized them in the following way : we assumed that the tide stream for a given coefficient is obtained by interpolation between the coefficients 45 and 95 ; we calculate the equivalent tide stream for coefficient 95.

The comparisons of radar current speed with the projection of the tide stream on the radar beam are shown in Figures 9, 10 and 11. Each point is one radar measurement. The dashed lines are the tide stream variations deduced from the tide tables. The crosses represent the mean and the standard deviation within one hour. The results are accurate for low ranges of tide. Results are also fairly good for great coefficients but a little different.Since the aim of the experiment was not the current measurement, there were a few measurements for those coefficients when the turbulence was very strong.

We may conclude that the mean current measurement is a good one.

This result is   important because it means that the Bragg lines are probably what we think they are, i.e. a representation of the current in a certain resolution cell of the radar we know, close to point 429. The results are very bad if we try to compare radar measurements in this resolution cell to other points of measurement of SHOM, even if they are not far away. So the degradation of the results on turbulence, which we will see in the next section, is not probable. These degradations should lead to a change in the mean current speed obtained by the radar.

5. TURBULENCE MEASUREMENTS : RESULTS

As we saw before, we characterize the turbulence with the standard deviation σ of the current distribution given by the shape of the Bragg lines. σ is calculated only with the most powerful Bragg line. Results for σ are not normalized by the tide range. So Figures 12 and 13 give absolute values of σ versus time in a tide period.

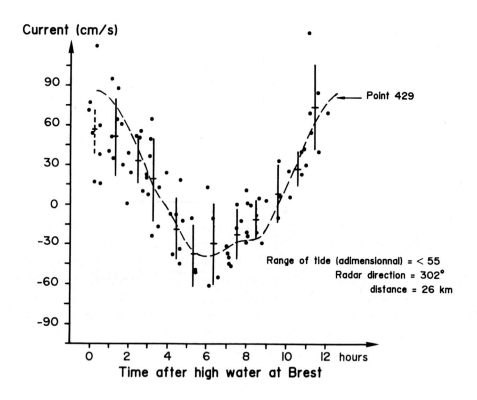

Fig. 9. Tidal stream component along the radar beam.
Dashed line : SHOM measurements
Points : OSUREM measurements with mean and standard deviation shown by the crosses.
Low ranges of tide.

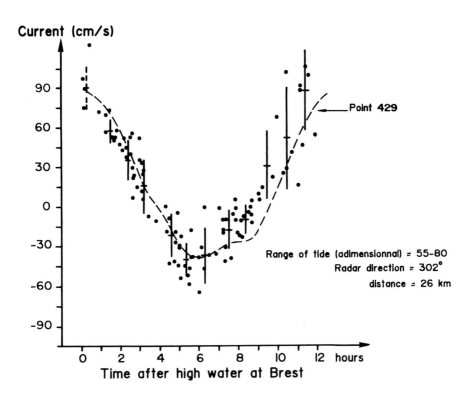

Fig. 10. Like Figure 9, but for middle ranges of tide.

270

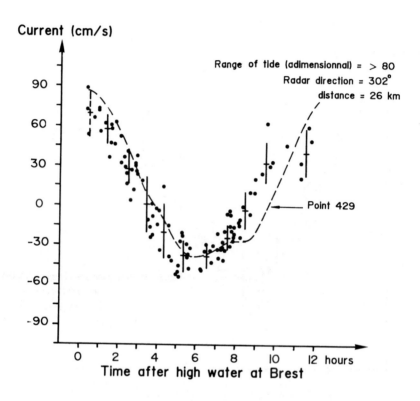

Fig. 11. Like Figure 9 and 10, but for high ranges of tide.

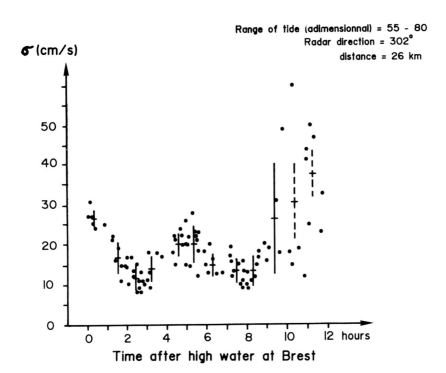

Fig. 12. Turbulence measurement for middle ranges of tide.

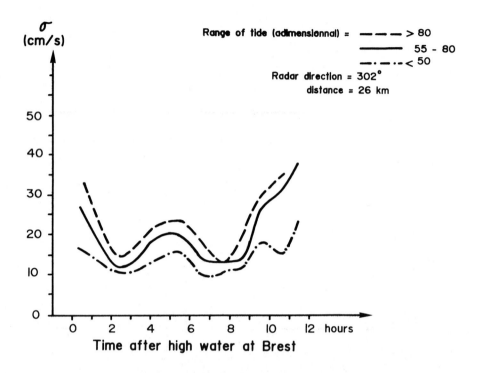

Fig. 13. Turbulence measurement. Dependence on the range of tide.

The turbulence is seen to be connected to the current velocity, i.e. to its variation in the semidiurnal tide period, and to the tide range.

When somebody measures the current versus time at a point in the sea, he considers as normal variations of this current of 10%. Here, this value leads to 13 cm/s for the maximum current with the maximum tide range (approximately the coefficient 110). The results obtained here are more than two times higher. The discussion of why the turbulence is so high in that area is not the purpose of this paper. We may suppose it is a characteristic of that area, but also that spatial measurement of the turbulence is not usual. Everybody knows that oceanic diffusion is often higher than diffusion calculated with simple models. So turbulent dispersion is usually calculated with empirical coefficients.

6. DISCUSSION

This section argues about the results to show that no artefact was found which would have been able to give such wide Bragg lines, except for turbulence.

Oceanographic remote sensing is a new science. The sensors used introduce new concepts, and it is not easy to connect their measurements with more conventional ones. It is always necessary to study their results carefully in order to eliminate eventual artefacts which may damage the results.

In this section we will review the possible phenomena which may produce the broadening of the Bragg lines.

6.1. The side lobes of the reception antenna

The reception antenna is composed of sixteen simple omnidirectional antennas. The first side lobe is - 16 dB below the main lobe. There is a natural protection against lobes at more than 90° from the radar beam by more than 10 km of land.

An attempt was made to study the importance of side lobes. We chose a period when the wind was up to the radar. In this case the effect of side lobes is below the limit of - 10 dB which is used. There is no difference between these results and those obtained in other wind directions. Another important item is that the mean current is effectively compared to the tide stream, and that even in upwind conditions, the mean current is the same for the negative and positive Bragg lines.

The shape of the Bragg lines is often flat like in Figure 4. Side lobes would have produced other shapes, without any flat central part.

Finally, there is no difference between the two frequencies 7 and 14 MHz. Perhaps the lobes are totally different.

So we may say that the results are not due to the radar characteristics and that all the signals came from the selected resolution cell.

6.2. The effect of the wind

We tried to find a correlation between the width of the Bragg lines and the wind speed in the area. The wind was often very low. Obviously no correlation was found in this case. But even with strong winds, we obtained the same result. We suppose that the wind is responsible for some dispersion in the results for the mean current. But simple corrections which were successful for the Marsen experiment [6], were not in this one.

Stokes drift is usually low, much lower than the results obtained for σ [9].

Witte [10] found that dispersion was a lot higher than that explained by the Stokes drift.

Since there is a strong wave-current interaction in some parts of the area closer to the coast [11], we studied its effect on the Doppler spectrum. In the case of crosswind conditions, the Bragg lines would be symmetrical about a frequency near zero. But the shapes observed are the effect of a 2 f_B translation of one Bragg line on the other, instead of a symmetry.

6.3   Other features

The σ value doesn't depend on the time scale. Measurements were made for 3 and 20 minuts. The results are already the same. So we are not measuring basically time-varying process. Since the results are independent of the radar frequency, they don't depend on the current variation with depth in the first meters. This is not surprising for a turbulence on the scale of kilometers.

Finally, there are a few variations of σ with the distance from the coast. But the farther we go the greater is the width   of the cell. So there is no definite conclusion on that point.

CONCLUSION

The HF narrow-beam radar may measure a turbulence characteristic, in addition to wind direction, long-wave directional spectra, total wave height and current speed. This turbulence value, which is the spatial standard deviation of the current speed, is not a classic one. Its values are surprisingly high off Ushant Island. It is well known that the wind effect on current in that area is very great... and difficult to calculate. We may assume that some variations of current with time have been supposed to proceed from the wind effect, and are in fact due to turbulence on the space scale of 5 to 10 km.

REFERENCES

[1] Broche, P., 1979. Sea state directional spectra observed by HF Doppler radar. Agard Conf. Proc., 263, 31.1-31.12.
[2] Parent, J. and Delloue, J., 1982. Determination de la direction du vent à la surface de la mer au moyen d'un radar à rétrodiffusion ionosphérique. Ann. Geophys., t. 38, fasc. 6, pp.863-873.
[3] Gay, H., Blanchet, C., Nicolas, J. and Piau, P., 1982. Determination of wind and short wave direction at great distances with OSUREM radar. In Wave and Wind Directionality. Ed. Technip, Paris.
[4] Forget, P., Broche, P., De Maistre, J.C. and Fontanel, A., 1981. Sea state frequency features observed by ground wave HF Doppler radar. Radio Science, Vol. 16, No 5, pp.917-925.
[5] Lipa, B. and Barrick D., 1982. Codar measurements of ocean surface parameters at ARSLOE. Preliminary results. Oceans 82.
[6] Janopaul, M.M. et al., 1982. Comparison of measurements of sea currents by HF radar and by conventional means. Int. J. Remote Sensing, vol. 3, No 4, pp. 409-422.
[7] Barrick, D. and Snider, J.B., 1977. The statistics of HF sea-echo Doppler spectra. I.E.E.E. Trans. on Antennas and Propagation, vol. AP-25, No 1.
[8] Service Hydrographique et Océanographique de la Marine, 1968. Tome No. 550. Courants de marée dans la Manche et sur les côtes françaises de l'Atlantique.
[9] Broche, P., de Maistre, J.C. and Forget, P., 1983. Mesure par radar décamétrique cohérent des courants superficiels engendrés par le vent. Oceanologica Acta, Vol. 6, n° 1.

[10] Witte, H. et al., 1982. Small scale dispersion measurements of drifter
     buoys in the North Sea. First int. cong. on meteorology and air/sea
     interaction of the coastal zone. The Hague, May 10-14.
[11] Cavanié, .A., Ezraty, R. and Gouillon, J.P., 1982. Tidal current modula-
     tions of wave directional spectra parameters measured with a pitch and
     roll buoy west of Ushant in winter, First international conference
     on meteorology and air/sea interaction of the coastal zone. The Hague.
     May 10-14.

A QUASI GEOSTROPHIC MODEL OF
THE CIRCULATION OF THE MEDITERRANEAN SEA

Laurent LOTH (*) and Michel CREPON (**)

(*) INRIA - Domaine de Voluceau - Rocquencourt - B.P. 105 - 78150 - LE CHESNAY - France

(**) Laboratoire d'Océanographie Physique - Museum National d'Histoire Naturelle - LA175 - CNRS - 43 Rue Cuvier - 75005 PARIS - France

Abstract

A quasi geostrophic model of the Mediterranean sea is solved by using a finite element method. The barotropic and baroclinic mode are computed independently. The Alboran Sea gyre is observed in both models but it is less intense than in nature. When penetrating the Mediterranean sea the Alboran sea current overshoots to the North, then becomes trapped by the Algerian shore.

1.INTRODUCTION

The Mediterranean Sea is a concentration basin. Evaporation creates a mass deficit in the whole basin which is compensated by an inflow of Atlantic water passing through the strait of Gibraltar and through the strait of Sardinia. The incoming Atlantic water which is light is tranformed into dense water by a complicated convective process (Gascard - 1978). This dense water forms a deep layer which flows out into the Atlantic ocean.These fluxes strongly influence the general circulation of the sea. From a schematic "point of view", the Mediterranean sea can be considered as a two layer ocean.

In the subsequent we focus our interest on the barotropic and baroclinic circulation of the western basin forced by the fluxes through the two straits.

2. THE MODEL

Since we are interested in low frequency phenomena we deal with the quasi geostrophic version of the shallow water equations.

The governing equations are, in a coordinate frame with x positive east-ward and y positive northward.

$$\frac{\partial}{\partial t}[\nabla^2 \Psi - \frac{1}{R^2}\Psi] + \mathcal{J}[\Psi, \nabla^2 \Psi - \frac{1}{R^2}\Psi] + \beta\frac{\partial \Psi}{\partial x}$$

$$= \frac{1}{\rho D} \text{ curl } \tau + A \nabla^4 \Psi - \varepsilon \nabla^2 \Psi \qquad (1)$$

where Ψ is the stream function ($u = -\frac{\partial \Psi}{\partial y}$, $v = \frac{\partial \Psi}{\partial x}$)

R is the internal Rossby radius of deformation.

($1/R^2$ is set equal to zero to obtain the barotropic mode)

β the variation rate of the Coriolis parameter ($\beta = 2.10^{-11} s^{-1} m^{-1}$)

ρ the density

D the depth

A the horizontal turbulent viscosity coefficient ($A = 512. m^2 s^{-1}$)

ε the bottom friction parameter ($\varepsilon = 5.10^{-7} s^{-1}$)

We only study the motion generated by fluxes of water flowing through the straits of Gibraltar and Sardinia. In this study the forcing due to the wind is neglected.

In order to satisfy the mass continuity (Pedlosky, 1979) it can be shown that

$$\frac{\partial}{\partial t} \iint_C \Psi \, dxdy = 0 \qquad (2)$$

Thus, to solve (1) subject to (2) we let (Holland, 1978)

$$\Psi = \Psi_0 + c(t) \Psi_1$$

where Ψ_1 is a solution of $\frac{\partial}{\partial t}(\nabla^2 - \frac{1}{R^2})\Psi_1 = 0$ with $\Psi_1 = 1$ on boundaries. Note that the time independent field Ψ_1 needs to be determined only once. The Stream function Ψ_0 is a solution of (1), with $\Psi_0 = \Psi_0^S = 0$ on the south boundary (African Coast) and $\Psi_0^S - \Psi_0^N$ is equal to the flux of water flowing through the strait, with $\Psi_0 = \Psi_0^N$ on the northern boundary (European coast). Now, condition (2) determines $c(t)$ at each instant, i-e.

$$c(t) = \frac{\iint_C \Psi_0 \, dxdy}{\iint \Psi_1 \, dxdy}$$

3. METHOD OF SOLUTION

In order to approximate the coastline geometry as closely as possible, we have chosen a numerical finite element approach, using a triangular grid (Fig. 1).

The model is forced by imposing velocity profiles at the two straits, the fluxes of which are equal. We start from rest at t = 0 and the two fluxes reach a constant value in one month. At each strait, the boundary condition is imposed at the end of a canal the length of which is four grid size. This allows the fluid to adjust itself before entering the sea and prevents unrealistic forcings in the basin.

The time discretization is a leap frog scheme with a Matsuno scheme every nine steps. The finite elements are interpolated by linear functions (Dumas - 1982, Dumas et al - 1982).

The bottom is assumed to be flat. The barotropic and baroclinic modes are solved separatly ($1/R^2$ = 0 for the barotropic mode in 1). This implies that baroclinic unstability is not taken into account. The depth of the upper layer is 200 m and the reduced gravity parameter g' is $10^{-2}ms^{-2}$ (g'=g $\Delta\rho$ /ρ) i.e the internal radius of deformation R is equal to 40 km. This value is larger than the actual one, but allows us to deal with a minimum number of triangles and to respect the dynamical constraints between the internal radius of deformation and the grid size which is taken about half of this value i.e. 20 km. At the coast a free slip condition is used.

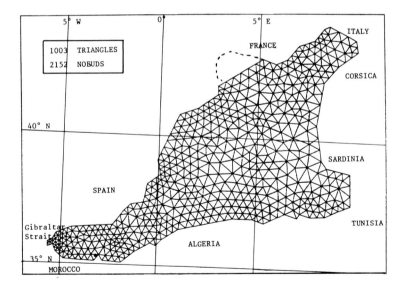

Fig. 1 : Finite elements grid used for the Mediterranean sea.

Many runs were performed in order to check the sensitivity of the model to the width of the strait of Gibraltar, to the velocity profile, to the magnitude of the incoming fluxe and to the eddy viscosity coefficient.

The final runs were done with realistic parameters. The width of the strait of Gibraltar (Sardinia) was 20 km (160 km). According to Lacombe and Richez - 1982 -, the incoming (and out-going) flux was 0.32 Sverdrup for the barotropic model and 1.6 Sverdrup for the baroclinic one. The viscosity coefficient A was 512 m^2s^{-1}. In the strait of Gibraltar the grid size is 10 km. This makes it possible to vary the velocity profile of the forcing. In the following runs assymmetric parabolic profiles are used (Fig. 3). Equilibrium is reached after 400 days in the barotropic case, after 700 days in the baroclinic case. The time step is 4 hours.

4. RESULTS

In both runs, the main current is deviated to the northern coast of the Alboran sea (Fig. 2 and 4). A part of this flow is recycled southward and forms one or two anticyclonic eddies. In the barotropic case one observes the formation of two weak anticyclonic eddies which are separated at the level of Cape Tres Forcas (Fig. 3). In the baroclinic case, there is one strong anticyclonic eddy which extends through the whole sea (Fig. 5, 6). When penetrating the Mediterranean sea, the current overshoots to North. This overshooting could be responsible for the front which extends between the Balearic Islands and Sardinia and which is often observed on infra-red satellite images (Fig. 7) (Deschamps et al - 1984). Then the current bends southward and flows along the Algerian coast.

The pattern of the circulation in the Alboran sea is strongly dependent on the vorticity of the forcing. If the vorticity is positive, the Alboran gyre is enhanced, if the vorticity is negative, the main current is not any more deviated to the coast of Spain but follows the coast of Morocco and the gyre disappears.

Following Holland (1978), several length scales of interest are defined
$$W_i = (u/_\beta)^{1/2} = 70 \text{ km}$$
$$W_s = \epsilon/\beta = 10 \text{ km}$$
$$W_m = 2(\frac{A}{\beta})^{1/3} = 60 \text{ km}$$
where u is a typical velocity (u = 0.1 ms^{-1})

The length scales W_i, W_s and W_m are respectively the width of the western boundary current when inertial effects dominate and when bottom friction dominates and when lateral friction dominates. These values show that the circulation in the Alboran sea is strongly dependent on inertial effect and lateral friction. Thus, an analysis of the motion in terms of vorticity balance must include the friction term.

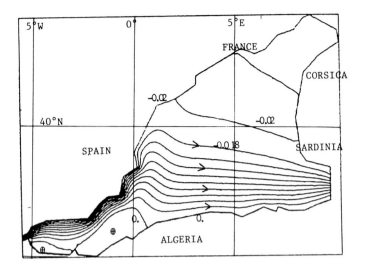

Fig.2. Stream lines of the barotropic model. The fluxes at the straits are equal to 0.32 Sverdrup. The distance between two stream lines is 0.032 Sverdrup.

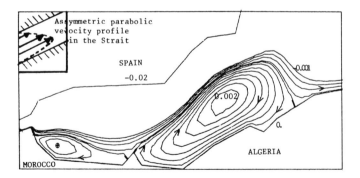

Fig.3. Barotropic model. Enhanced picture of the stream lines in the Alboran sea. The distance between two stream lines is 0.003 Sverdrup. The velocity profile of the forcing in the Gibraltar strait is shown in the upper left corner.

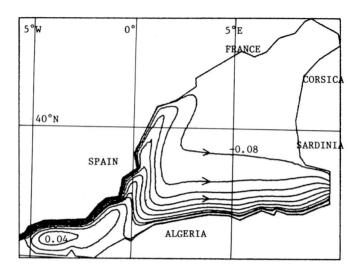

Fig.4. Stream lines of the baroclinic model. The fluxes at the straits are equal to 1.6 Sverdrup. The distance between two stream lines is 0.16 Sverdrup.

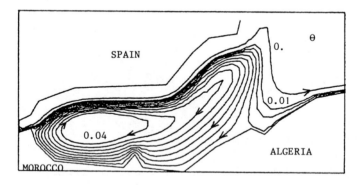

Fig.5. Baroclinic model. Enhanced picture of the stream lines in the Alboran sea. The distance between two stream lines is 0.05 Sverdrup. The velocity profile of the forcing is the same as in Fig.3.

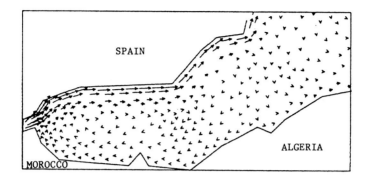

Fig.6. Baroclinic model. Velocity vectors in the Alboran sea.

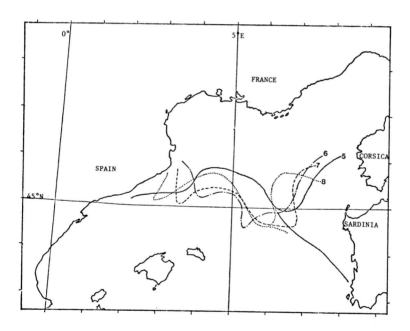

Fig.7. Pattern of the thermal front between the Balearic Islands and Sardinia at different months of the year 1978 observed from NOAA satellite (from Deschamps et al. 1984) - (5 is May, 6 June, 7 July, 8 August).

5. CONCLUSION

The finite element technics is a valuable tool to study the mediterranean circulation. The model supports realistic inflows of Atlantic water passing through the strait of Gibraltar. Despite the over-simplification of the model, many features of the circulation as the Alboran gyre (Lanoix - 1974) are reproduced. But it is noted that the circulation in the northern part of the Basin is not obtained. In particular, the strong cyclonic gyre (Crepon et al. 1982) existing between France and Corsica is not observed. The next stage is to include the wind stress and the bottom topography by dealing with a two layer model which can trigger the Alboran sea gyre more intensely.

The computations were done on the Cray 1 of French Research.

ACKNOWLEDGEMENTS

This work was supported by french contrat Dret N°81/1117 and by CNRS and CNEXO. The finite element model was kindly provided by C. Le Provost. Discussion with P. Delecluse and J.C. Gascard, C. Millot have been very helpfull throughout this work.

REFERENCES

Crepon M., Wald L. and Monget J.M. - 1982 - Low frequency waves in the Ligurian sea during December 1977. J.G.R Vol. 82 C1 pp 595-600.

Deschamps P.Y., Frouin R. and Crepon M. - 1984 - Sea surface temperature of the coastal zones of France observed by the HCMM satellite - J.G.R (in Press)

Dumas E. - 1982 - Modélisation des circulations océaniques générales par des méthodes aux éléments finis. Thèse - University of Grenoble - June 1982.

Dumas E., Le Provost C. and Poncet A. - 1982 - Feasibility of finite element methods for oceanic general circulation modeling. In Proc. of 4th. Int. Conf. on finite elements in Water Res. HANOVER - 1982.

Gascard J.C. - 1978 - Mediterranean deep water formation ; baroclinic instability and ocean eddies - Oceanologica Acta - Vol. 1, N°3 - pp. 315-330.

Holland B. - 1978 - The role of mesoscale eddies in the general circulation of the ocean; numerical Experiment Using a wind-driven quasi geostrophic model. J.P.O. Vol. 8 N°3 pp. 363-392.

Lacombe H. and Richez C. - 1982 - The regime of the strait of Gibraltar in hydrodynamics of semi-enclosed seas. In hydrodynamics of semi-enclosed seas by J.C.J Nihoul (Editor), Elsevier, Amsterdam, pp. 13-73.

Lanoix F. - 1974 - Project Alboran - Etude hydrologique et dynamique de la mer d'Alboran - NATO technical report 66, 39 p.

Pedlosky J. - 1979 - Geophysical Fluid Dynamics - Springer Verlag - 624 p.

SOME APPLICATIONS OF REMOTE SENSING TO STUDIES IN THE BAY OF
BISCAY, CELTIC SEA AND ENGLISH CHANNEL

R.D. PINGREE

Institute of Oceanographic Sciences, Wormley, Surrey, GU8 5UB,
England

ABSTRACT

Infra-red, Coastal Zone Colour Scanner and Synthetic Aperture
Radar images have been used to identify sea surface structures in
the Biscay, Celtic Sea and English Channel regions. Attention
has been focussed on shelf-break cooling, shelf-break chlorophyll
'a', Biscay eddies, internal waves and turbidity structures in
the English Channel and extended where possible with examples
drawn from work at sea.

INTRODUCTION

One of the most important contributions of remote sensing to
oceanographic and shelf studies in the Biscay, Celtic Sea and
English Channel has been to provide clear illustrations of a
variety of physical and biological phenomena. With a clear
picture in mind of the process and its geographical limits it
becomes a relatively simple matter to investigate the processes
further with measurements at sea. For example there were no
reports of the extensive shelf-break cooling in this area until it
had been first noted in the infra-red satellite imagery. The
widespread occurrence and persistence of the shelf-break cooling
stimulated models of both the M_2 barotropic tidal currents and
the internal tide for this area. In this paper some examples of
the kinds of structures that can be observed using remote sensing
techniques are drawn from the Biscay, Celtic Sea and English
Channel and illustrate shelf-break cooling and associated
phytoplankton blooms, Biscay eddies, shelf tidal fronts, coastal
upwelling, frontal eddies and instabilities, internal waves and
turbidity structures. Some supporting sea truth is also presented
but it is clear that realistic models for these processes is a
subject of future research.

Fig. 1. Infra-red satellite image (1339 GMT, 26 August 1981) illustrating shelf-break cooling. The Ushant, Scilly Isles, Lands End, and Celtic Sea tidal fronts can also be identified. In addition, coastal tidal fronts occur along the Armorican shelf. Cool water due to previous wind induced upwelling also occurs off Southwest Ireland and the Spanish Coast. High pressure, calm wind conditions existed on 26 August 1981 and sea surface temperature 'hot-spots' (see Fig. 4) associated with windless high pressure conditions can be seen in the Celtic Sea and western English Channel.

1. SURFACE TEMPERATURE STRUCTURE

1.1. <u>Shelf Tidal fronts, banded structures and upwelling fronts</u>

Infra-red satellite imagery has revealed clearly the tidal
fronts in the English Channel, Celtic Sea and Armorican shelf
(Figs. 1 and 2). These transitions between tidally mixed and
stratified water (Simpson and Hunter, 1974; Pingree and Griffiths,
1978) persist for about ∿ 100 days over the summer months June,
July, August. The boundaries between mixed and stratified waters
appear to be unstable and are characterised by irregular eddies

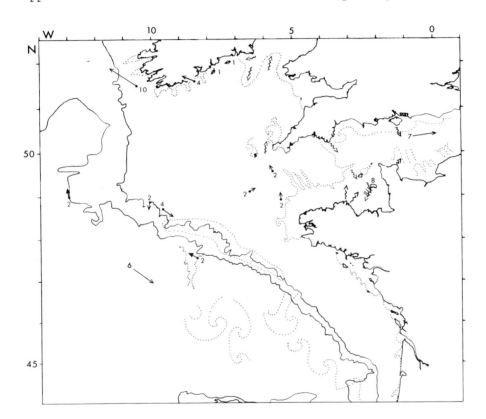

Fig. 2. A sketch of some features observed in the infra-red
satellite imagery (dotted lines). Also shown are some current
measurements (see text for explanations). A dot signifies the
position of the current meter mooring and on the shelf only the
measurements refer to the upper part of the water column. A wavy
arrow represents flow inferred from satellite images. An arrow
without a dot indicates the movement of a surface drifting buoy.
The numbers used give the mean speed in cm s^{-1}. Also shown are
the 100 fm and 1000 fm contours.

which occasionally show a tendency to be cyclonic (Pingree et al.,
1979), as exemplified by the Lands End frontal zone (Fig. 1). The
instabilities have time scales of order 1 day and length scales of
about 20 km. Long (30 km) intrusive fingers with separations of
10-20 km can also be observed on the Ushant front and these
features tend to be particularly conspicuous in September as the
mixed region increases in area and extends across the mouth of the
English Channel. At this time of year the Ushant tidal front can
show a marked spring-neap variation in geographical extent.
Frontal instabilities are thought to represent an important agency
in the cross frontal transfer of water properties.

Two interesting features that have also shown up with infra-red
imagery (which still require sea truth to confirm that they are
indeed real features of the sea surface temperatures rather than
atmospheric effects) are the bands of apparently cold water that
appear on-shelf extending approximately normal to the shelf-break
in May and June and parallel to the shelf-break in July and August.
The normal bands extend for $\sim$ 100-200 km and have a wavelength $\sim$
15 km and probably result from mixing or internal tides associated
with the linear tidal sand ridges that occur in the Celtic Sea on
similar scales $\sim$ 15 km, fig. 3(a). The parallel bands have
wavelengths of 20-30 km and are also relatively stationary. It
might be argued that they occur where the currents of the
barotropic tide are in phase with the currents of a progressive
internal tide resulting in increased local mixing.

Infra-red satellite imagery has also identified clearly the
marked seasonal upwelling that occurs along the Spanish and
Portuguese coast and shown that upwelling also occurs occasionally
off Southwest Ireland and on that part of the French coast which
is adjacent to the Armorican and Aquitaine shelf.

1.2. Shelf Break cooling and Biscay mesoscale eddies

Whilst some of the gross features listed above were known
before the widespread use of infra-red satellite imagery there
were few reports of shelf-break cooling and none showing the
characteristically deep Biscay eddy structure. Shelf-break cooling
extends typically for 300 km along the shelf-break and slope
regions and persists from late May to late September and like the
tidal fronts is characterised by irregular smaller scale
structures. Shelf-break cooling has not been observed in winter

Fig. 3(a). Infra-red satellite image (27.5.83) showing apparently cool bands extending approximately normal to the shelf-break and associated with the sand ridges which have similar transverse wavelengths ∿ 10-15 km. The whiter features are clouds and should be ignored. The island on the lower right is Ushant and the start of the Ushant front is indicated by the wispy features stretching from the French coast. The pass corresponds to spring tides and high pressure atmospheric conditions and sea mist may perhaps serve to accentuate some of these features.

though shelf-break warming has been noted in January 1979 and 1982 (as far north as 47°N) and presumably associated with advection and spreading of warm water from the Spanish coast. Shelf-break cooling has been attributed to result from mixing by internal tides simply because its position corresponds approximately to the region where the M_2 tidal currents have maximum values and this aspect is discussed in more detail later. However upwelling processes, mixing by trapped waves and inertial currents are all thought to play a contributing role in shelf-break cooling. Cooler water from the slopes has been observed spreading onto the shelf for limited distances (∿ 15 km) from Penmarch, Guilvinec and Odel canyons.

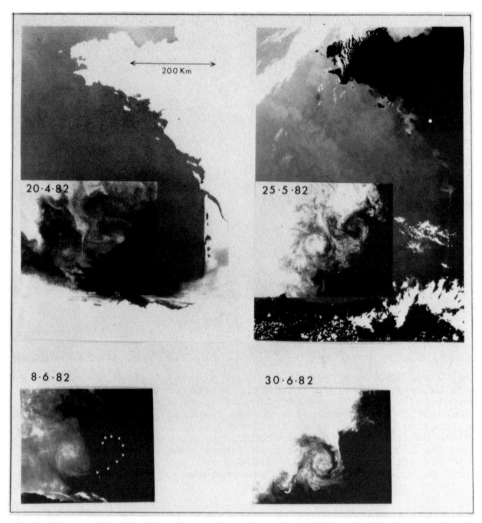

Fig. 3b.   Infra-red satellite image showing an evolving Biscay vortex pair. The lower images (8.6.82 and 30.6.82) fit into the Biscay region in the same manner as the upper images (20.4.82 and 25.5.82).

In common with most oceanic regions, the deep Biscay shows large (100 km) interconnected eddies that are generally confined to the abyssal plain by the slopes. Individual examples of such eddies from infra-red images have been given by Frouin (1981) and Dickson and Hughes (1981). At sea they have been studied using

drifting buoys (Madelain and Kerut, 1978) and in the Tourbillon
experiment (Le Groupe Tourbillon, 1983). They have the
characteristic structure sketched in Fig. 2 and can last for a
week or more. The anticyclonic structure of the vortex pair
illustrated in Fig.3(b) appeared to persist in one form or another
over a period of two months apparently fed by (or drawing in) a
cool stream of water flowing along the base of the slopes.
Although this vortex pair appears in the central Biscay from the
point of view of the 3000-4000 m topography it is pressed up
against the base of the slope in the S.E. corner of the Biscay
and seems to subsequently put some cooler water up on the Spanish
slope. Biscay eddies also appear to be able to draw water off
the shelf and cool plumes can sometimes be observed extending
from the region of shelf-break cooling for distances of 100-200km,
Fig. 2.

In addition to the interconnecting vortex pair structures,
cyclonic eddies with wave lengths of 100 km have been observed in
the S.E. Biscay which appear to be confined to the lower part of
the slopes and are also sketched in Fig. 2.

1.3.   Hot spots
Cloud-free infra-red satellite images generally occur under
high atmospheric pressure conditions. The daytime passes then
show "hot-spots" as the sea surface warms up in localised places
where conditions are relatively windless. Measurements at sea
and from drifting buoys have shown that under such windless
conditions (< Force 2) the top metre can warm up by 2-3°C
(Fig. 4). Such effects have allowed structures to be observed in
the generally tidal mixed conditions of the central and eastern
regions of the English Channel, for example, fresher water
spreading from the Bay of Seine region particularly at neap tides,
and the effects of tidal mixing and topography in the Channel
Isles area.

2.   SURFACE CHLOROPHYLL STRUCTURES
2.1.   Shelf-break frontal Fluorescence and Reflectance
Measurements at sea have shown that both the Ushant front and
the shelf-break cooling region can show increases of chlorophyll
'a' at the surface (Pingree et al., 1982). This is thought to be

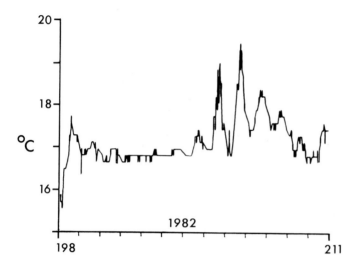

Fig. 4. Surface temperature record from surface drifting buoy (which followed the 2000 m contour northwestward in the vicinity of 9°W at about 5 cms⁻¹) showing marked diurnal temperature variations.

due to the favourable nutrient and light regime afforded by the physical mixing processes. In June a band of inorganic nutrients occurs along the shelf-break with nitrate values typically ∿ 1 μg at l⁻¹ N-NO₃ (Fig. 5). In July, August isolated, higher than background, nitrate-nitrogen patches occur with generally cooler water showing that there is, indeed, on occasions, a nitrate source at the surface that can be utilised by phytoplankton phytoplankton growing near the surface (Fig. 6). Whilst the values of fluorescence at the shelf-break are very variable with exceptionally high values associated with some nannoplankton communities (for example the Prasinophycean flagellate Micromonas sp (1-2μ diameter) together with the Chrysophycean flagellate Pseudopedinella sp (6μ diameter)), the chlorophyll 'a' values are typically only ∿ 1 mg chl'a' m⁻³, an order of magnitude lower than the values that are commonly associated with blooms in the vicinity of the shelf-tidal fronts or which occur during the spring bloom in the Celtic Sea. However mackerel eggs occur in maximum number at the shelf-break in May-June (Coombs et al., 1981) and it may be the large geographical extent of this region of increased levels of surface chlorophyll 'a' and the associated

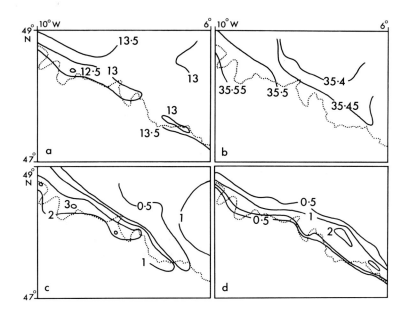

Fig. 5. (a) Surface temperature (°C);   (b) salinity (°/$_{oo}$);
(c) chlorophyll 'a' (mg m^{-3}) and (d) inorganic nitrate (μM)
(3-6 June 1983).   200 m contour shown by dotted line.

protracted productive season of both primary and secondary
production which provide the ecological advantages that favour
this spawning area.

The Coastal Zone Colour Scanner (C.Z.C.S.) imagery has shown
more clearly than ever before the geographic scale and
persistence of the shelf-break blooms (Fig. 7).   Chlorophyll
absorbs more strongly at the blue end of the visible spectrum than
in the yellow part and in broad terms a measure of the chlorophyll
from C.Z.C.S. data can be obtained from the ratio of the
reflectances from channel 1 (blue, 443 nm) or channel 2 (green,
520 nm) to channel 3 (yellow, 550 nm) after applying an
atmospheric correction to each using channel 4 (red, 670 nm).

Some of the spectacular blooms that have been observed at the
shelf-break are comprised mainly of coccolithophores (Holligan
et al., 1983) which give a characteristic milky appearance to the
water.   The calcite plates of the coccolithophores are strongly
reflecting and the structure of these blooms can be seen in the
raw channel 3 data.   In common with the infra-red imagery the
C.Z.C.S. imagery has shown thin plumes extending out from the

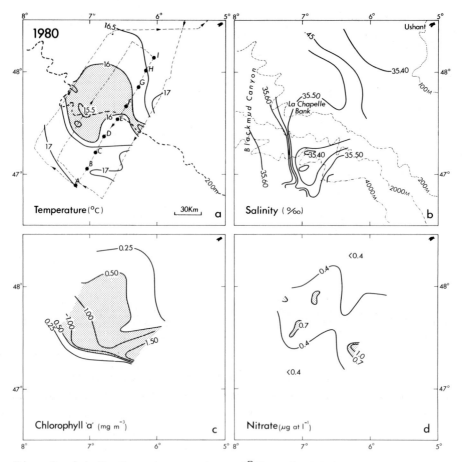

Fig. 6. (a) Surface temperature ($^{\circ}$C) and ship's track;
(b) salinity ($^{\circ}/_{\circ\circ}$); (c) chlorophyll 'a' (mg m^{-3}) and
(d) inorganic nitrate (μM) (August 1980). Bottom topography
is given in metres.

shelf-break and eddies in the deeper Biscay regions. The plumes
of phytoplankton drawn off from the shelf-break slope region are
further evidence of physical processes (in this case Biscay
eddies) and may be important in the development and subsequent
decay of shelf-break blooms.

2.2.   Fluorescence Along Shelf tidal fronts
    Increases in reflectance also occur along the Ushant tidal
front (Fig. 7) where physical mixing processes again control the

Fig. 7. C.Z.C.S. (Coastal Zone Colour Scanner) image (22 June 1981) showing regions of relatively high surface chlorophyll in the vicinity of the shelf-break and to the stratified side of the Ushant front.

availability of nutrients and light. As the season progresses the phytoplankton composition changes from a dominance of diatoms to a dominance of dinoflagellates which tend to occur in the stratified waters adjacent to the tidal fronts. Spectacular surface blooms of dinoflagellates (Gyrodinium aureolum) have been observed in 1975, 1976, 1978, 1981 which extend from the frontal boundary well across into surface waters of the shallow thermocline ($\sim$ 20 m) regions of the western English Channel where values of chlorophyll 'a' as high as $\sim$ 100 mg chl'a' m^{-3} have been recorded. The precise role of water movement, nutrient fluxes and vertical migration of the dinoflagellates in maintaining these

surface distributions in the shallow stratified waters adjacent
to the Ushant tidal front is a subject of continuing research.

3.   INTERNAL WAVES AND TIDES

(i)   Surface radar structures. Whilst shelf-break cooling may
be considered as possible indirect evidence for internal tides,
infra-red satellite imagery has not yet provided clear examples of
internal tides. This is hardly surprising since measurements at
the shelf-break near 48°N have shown that although the
thermocline may oscillate by more than 50 m at spring tides
(Fig. 8) there may be no surface temperature expression of the
internal tide. The synthetic aperture radar (S.A.R.) on board the
SEASAT on the other hand has provided striking examples of
internal waves in the Biscay region and allowed an estimate to be
made for the phase speed for the internal tide propagating

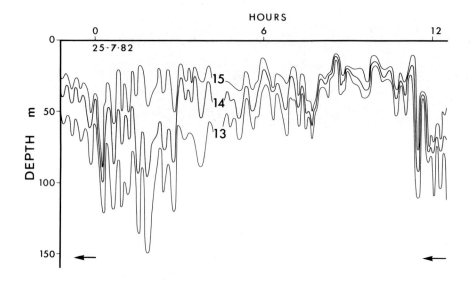

Fig. 8.   Isotherms (°C) obtained from repeated S.T.D. profiles
(every 10 mins) near the shelf-break in about 250–350 m depth
following a drifting dahn whose approximate mean position was
48°08'N 8°11'W. The higher frequency oscillations of about
15 min period could be contoured without aliasing using the echo
sounder to monitor the acoustic scattering layers in the
thermocline. The time of maximum off-shelf tidal streaming is
indicated by an arrow. The tidal oscillations of the thermocline
(14°C contour) in this region have a peak to trough displacement
of about 50 m at spring tides. There is also a marked second
barocline mode. The 16°C contour near the surface is not shown.

on-shelf. Variations in sea surface roughness due to the
interaction of the internal tidal currents with the surface waves
allows the radar to reveal the internal waves clearly (Fig. 9).
Such waves can, in fact, be observed at sea using ship's radar
(see for example Haury et al., 1983) or even visually as a result
of the increased number of breaking waves associated with the
internal wave trains. On occasions parallel "walls of white"
water (breaking surface waves) separated by about 1 km can be seen
stretching for several miles indicating the presence of large
internal waves propagating on-shelf.

The S.A.R. image for this region (Pingree and Mardell, 1981)
indicates that although there are many sources for the internal
waves they mainly originate at the shelf-break from localised
sources. Internal waves appear to move on-shelf about $\sim$ 30 km in
what is assumed to be a tidal period giving a phase speed of
67 cm s^{-1}. They also propagate off-slope and out into the Biscay.

The internal waves illustrated in Fig. 9 have wavelengths of
order 1 km and non-linear effects are important in their
generation and subsequent propagation. Such images have
stimulated the development of numerical models and measurements
of the internal tide using thermistor chains and current meter
moorings. Since the internal tides are thought to represent one
of the main candidates causing shelf-break cooling and the
associated shelf-break fluorescence it is of interest to consider
some of the possible characteristics of the internal tides in
this region.

(ii) Numerical models

The following simple numerical model neglects rotation,
assumes the shelf-slope region has a regular geometry and is only
valid for long waves (so lee wave formation where non-hydrostatic
pressure becomes important is not taken into account). Although
in its present form the model may not be very realistic it does
show that long wave internal tides might be forced by the
barotropic tide as the tidal currents move up and down the slope
thereby causing oscillations of the thermocline. In this model
a crest is formed near the shelf-break just after on-shelf tidal
streaming, whereas a trough forms just after off-shelf tidal
streaming. The crests and troughs divide in the slope region
near the shelf-break and propagate as free waves both on-shelf
and off-shelf towards the ocean. Since the trough formed during

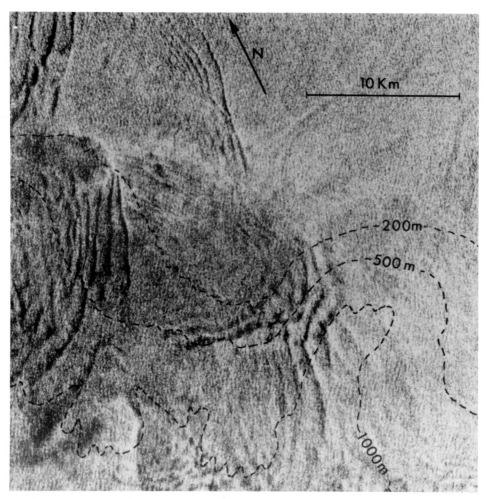

Fig. 9. A digitally processed portion of the synthetic aperture radar (S.A.R.) pass on 20 August 1978 showing internal waves in the shelf-break region with wavelengths typically of order 1 km. The image centre is located at 46°51'36"N, 5°9'58"W. The pass corresponds to tidal conditions 1 hour after maximum off-shelf tidal streaming at spring tides.

off-shelf tidal streaming is propagating on-shelf against the tidal current and a marked steepening of the internal wave profile occurs which propagates as an internal tidal bore.

The model is a vertical section normal to the shelf-break spanning oceanic (4000 m), slope and shelf regions (200 m). The thermocline is represented by an upper layer h' of density ρ' and a lower layer, h", of density $\rho"$. The x axis is chosen positive

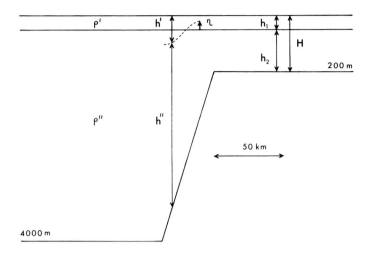

Fig. 10. Schematic representation of the model showing thermocline spanning oceanic, slope and shelf regions. The grid scale is 500m.

in the on-shelf direction and the grid scale is 500 m (Fig. 10). For simplicity conditions are taken as uniform in the along-shelf sense. The internal tide U is defined by U = u' - u" where u' is the current in the upper layer and u" is the current in the lower layer. The barotropic tide, or vertically integrated tidal current, u, is assumed to be unmodified by the internal tide and is specified in advance in accordance with shelf-slope geometry.

The equation of continuity for the upper layer can be transformed into an equation for the internal oscillation η against time t, to give

$$\frac{\partial}{\partial x} (h'u) + \frac{\partial}{\partial x} (\frac{h'h''}{H} U) = \frac{\partial \eta}{\partial t} \tag{1}$$

where $H = h' + h'' = h_1 + h_2$ and $uH = u'h' + u''h''$

The first term on the left hand side of equation (1) is the source term for the internal displacement of the thermocline η. It also allows the barotropic tide to move the internal tide back and forth on the shelf since $h' = h_1 - \eta$. Variations in surface elevation are neglected with respect to the internal oscillation.

The barotropic tide, u, is prescribed according to the non-divergent equation

$$\frac{\partial}{\partial x} (Hu) = 0 \tag{2}$$

Thus the source term for the internal tide varies as $-(^1/_H)^2 \frac{\partial H}{\partial x}$

and has a maximum value just at the top of the slopes when

$\frac{\partial H}{\partial x}$ is constant.

A simplified momentum equation is obtained by subtracting the momentum equations for the upper and lower layers to give

$$\frac{\partial U}{\partial t} + \frac{\partial}{\partial x} (uU) = \beta \frac{\partial \eta}{\partial x} + K\nabla^2 U \tag{3}$$

where $\beta = g(1 - \frac{\rho'}{\rho''})$ is the reduced gravity and g is the acceleration due to gravity.

More complete forms for the term arising from advection $\frac{\partial}{\partial x} (uU)$ gave qualitatively similar results and in this simple treatment are not further discussed. The term $K\nabla^2 U$ represents attenuation by diffusion, with coefficient K, and also assists with numerical stability.

The mean depth of the upper layer was taken as $h_1 = 30$ m and the slope region had a uniform gradient of 1 in 10 from $H = 200$ m to $H = 4000$ m in 38 km. Thus in a linear model the phase speed is

$c \sim \left[\beta \frac{h_1 h_2}{H} \right]^{\frac{1}{2}} \sim 50$ cm s^{-1}, with $\beta \sim 1$ cm sec^{-2}, and $h_2 = 170$ m,

considerably less than that suggested by the S.A.R. image. The corresponding wavelength $\lambda = \frac{2\pi}{K}$ for M_2 tidal frequency, $\sigma = \frac{2\pi}{T}$ is $\lambda = 23$ km. K was chosen such that $Kk^2 \sim 2/\tau$ so free waves in a linear model would decay to 1/e of their amplitude after a time τ and τ was set $\tau = 4T$ where T is the M_2 tidal period, thus $K \sim 1.4 \times 10^2 m^2 s^{-1}$. The amplitude of the oscillating barotropic tide was taken as 75 cm s^{-1} which represents a peak spring tide condition for a ~ 100 km stretch along the shelf-break in the Celtic Sea Armorican Shelf region.

(iii) <u>Long waves without rotation</u>. The structure of the internal tide is illustrated by hourly sequences of the displacement of the thermocline. The linear model where all non-linear terms were neglected is shown in Fig. 11 and the results of the non-linear model using the full equations (1), (2) and (3) is shown in Fig. 12. In both models a trough occurs at the shelf-break just after maximum off-shelf tidal streaming.

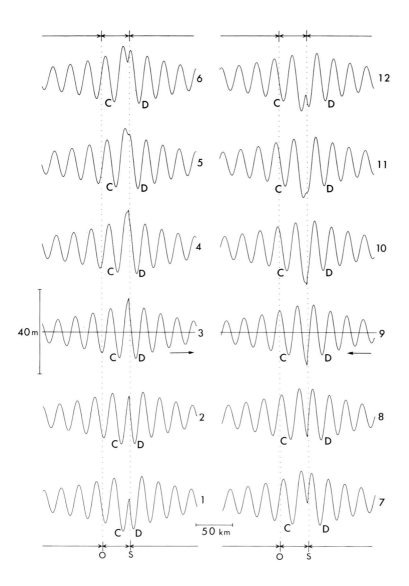

Fig. 11. Internal tidal displacements of the thermocline every lunar hour using linearised equations. The slope region extends from O to S where S is the shelf-break (200 m) and O is the start of the oceanic region (4000 m). D is an on-shelf propagating trough and C is an ocean-going trough. A vertical scale of 40 m is shown at hour 3. Hour 3 corresponds with maximum on-shelf tidal streaming and maximum off-shelf tidal streaming occurs at hour 9 (depicted by arrow).

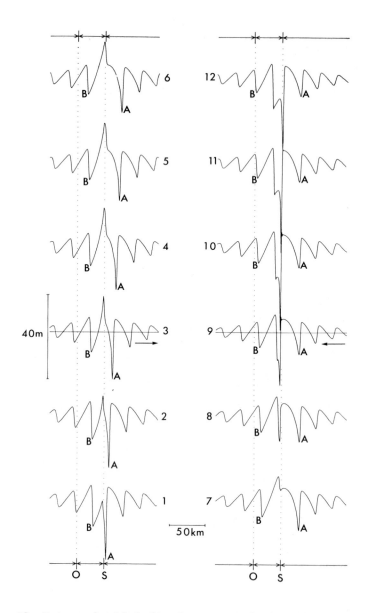

Fig. 12. Internal tidal displacement of the thermocline every lunar hour using the fully non-linear equations. The slope region extends from O to S where S is the shelf-break (200 m) and O is the start of the oceanic region (4000 m). B is an ocean going trough and A is an on-shelf propagating trough. Hour 3 corresponds with maximum on-shelf tidal streaming.

However in the non-linear model the leading edge of the on-shelf propagating trough is unable to move on-shelf against the barotropic tidal currents until the tidal streams slacken. This results in a very distorted and steepened trough for the internal tide which subsequently propagates rapidly across the shelf when the tidal streams are on-shelf and is halted and momentarily reversed in direction during maximum off-shelf tidal streaming.

(iv) Effect due to rotation. When rotation is taken into account and conditions are again uniform in the along-slope sense rotation increases the propagation speed. In addition relatively more energy is associated with the currents rather than the internal displacements of the thermocline. Linear theory and numerical model give the propagation speed for long waves with horizontal crests as

$$c^2 = g(1 - \frac{\rho'}{\rho''})(\frac{h_1 h_2)}{h_1 + h_2})(1 - \frac{f^2}{\sigma^2})^{-1} \tag{4}$$

where f is the Coriolis parameter and σ is the tidal frequency. For $f/\sigma \sim 0.77$ appropriate for these latitudes, this will result in an increase in phase speed and wavelength for the progressive internal tide of about x 1.6.

The waves are now dispersive and the group velocity, c_g, for linear long waves is defined as

$$c_g = \frac{\partial \sigma}{\partial k} = c(1 - f^2/\sigma^2) \tag{5}$$

and with $f/\sigma \sim 0.77$, as before, this gives a group velocity of about 1.6 times smaller than the phase speed of waves in the absence of rotation. This implies that any modulation of the long-wave internal tidal signal at the shelf-break due to the spring-neap cycle of the barotropic tide will travel only slowly on shelf or off shelf.

When the non-linear terms in equations (1) and (3) were included the distortions of the internal tide on the shelf were no longer as indicated in Fig. 12 but had deeply penetrating troughs and the wave profile on the shelf also tended to be symmetric with respect to the troughs.

(v) Shorter wavelengths. The waves so far considered assume that the internal tidal currents are uniform in top and bottom layers. Clearly this is not valid for shorter wavelengths as

exemplified by the S.A.R. image which indicates non-linear
internal wave trains or packets of internal solitons.   More
realistic models would have to make allowance for the vertical
structure of the currents in the upper and lower layer for the
higher wave numbers as in the Korteweg and de Vries (1895) first
approximation.   Solitary waves and solitons travel at speeds in
excess of that given by small amplitude linear theory.   Their
fractional increase in phase speed is very approximately $\sim \frac{1}{2} \eta/h$,
(Alpers and Salusti, 1983) and so the finite amplitude of shorter
waves may produce increases in phase speed that could match the
value (~ 67 cm s^{-1}) inferred from the S.A.R. image.   Thus non-
linear effects of finite amplitude for the shorter waves or
the linear effects of rotation for the longer waves significantly
increases the propagation speeds of the internal waves.

(vi)   Measurements at sea.   An extreme example of the
structure of the internal tide at spring tides obtained from a
thermistor chain mooring placed on the shelf in the region of
maximum M_2 tidal currents at 47°40.0'N 6°19.1'W (about 20 km
from the shelf-break) is shown in Fig. 13 .   At spring tides the
barotropic tidal currents reach almost 2 knot at this position
and although the tidal currents are reduced at the shelf-break
they are still comparable with the phase speed of the internal
tide.   The trough formed during off-shelf tidal streaming is thus
propagating on-shelf against the tidal current and at spring
tides this will result in a marked steepening of the internal
tide.

Measurement made from fixed moorings will need correcting for
the distortions that occur as the tidal currents advect the
internal tide past the mooring.   Current measurements made near
the thermistor chain mooring showed that the leading edge of the
trough of the internal tide passed the thermistor chain mooring
when the on-shelf tidal current was about 1.5 knot (about 1.0
hours after maximum on-shelf tidal streaming).   Thus some of the
steepening associated with the trough of the internal tide is
apparent and due to making measurements at a fixed point rather
than following the oscillating barotropic tidal flow.   The
14°-15°C isotherms descend below 50 m for about 20% of the wave
period.   During this time a local water column would move
on-shelf about 1-2 km which is only a small fraction ($\sim 5\%$) of
the wavelength of the internal tide (~ 30 km).   So it appears

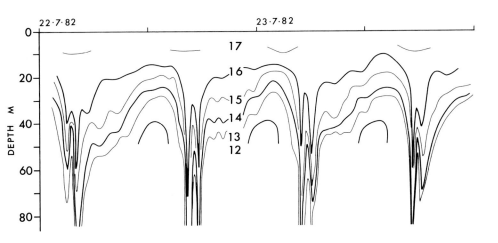

Fig. 13. Isotherms ($^{\circ}$C) from the thermistor chain mooring 069 (47°41.8'N 6°18.2'W). The measured structure of the internal tide propagating on-shelf is highly distorted with deeply penetrating troughs. There is also a noticeable second baroclinic mode. Some smoothing of the data was necessary to produce a clearer illustration. The period illustrated corresponds to spring-tide conditions with semi-diurnal currents typically ∿ 80-90 cm s^{-1} (vertically integrated).

that at spring tides, at least, the internal tide is distorted in this particular region with more deeply penetrating troughs.

A closer inspection of Fig. 13 shows that the troughs are generally composed of two large amplitude waves at this site. At some places near the shelf-break the internal tidal signal takes on the form of a group of short wavelength (∿ < 1 km) internal waves propagating on-shelf. An example is illustrated in Fig. 14 which shows a large amplitude wave followed by smaller waves and such waves are believed to cause the surface features seen in the S.A.R. image (Fig. 9).

Current measurements have also been made at the shelf-break near 47°30'N to see whether the shear produced by the internal tide is sufficient to cause mixing in the thermocline and contribute to the shelf-break cooling observed in the infra-red satellite imagery. Gradient Richardson numbers of ∿ 1 have been measured by current meters separated vertically by ∿ 84 m with temperature differences across the base of the thermocline of ∿ 1°C for periods of ∿ 1 hour at spring tides. It is hard not to draw the conclusion that a closer separation of current meters

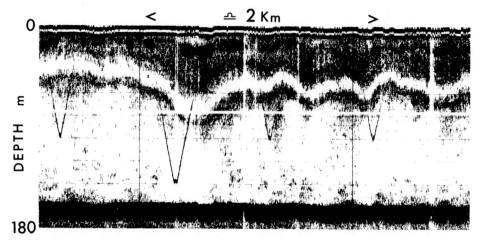

Fig. 14. Echo sounder trace from 47°52.5'N 6°29'W (20 n.m from the shelf-break (200 m contour) on 27.7.83) showing large amplitude waves on the thermocline propagating on-shelf which produce the features seen in the S.A.R. image (Fig. 9). (Near vertical lines show C.T.D. dips).

would produce even lower Richardson number values. C.T.D. profiles have also shown small-scale (< 5 m) instabilities in density in the seasonal thermocline in this general region.

As yet no clear spring-neap cycle of shelf-break cooling has been detected in the infra satellite imagery or established from measurements at sea and it seems likely that additional factors are involved. Maze (1983) has proposed that the combination of internal tides and inertial currents produced by wind events is a significant factor in shelf-break cooling.

4.   WATER MOVEMENTS INFERRED FROM TEMPERATURE AND TURBIDITY

4.1.   Infra-red

In general, infra-red and C.Z.C.S. satellite imagery can only give a rather qualitative indication of water movement in the Bay of Biscay and English Channel area. Whilst it is clear that the meso-scale eddies (∿ 100 km) in the deep Biscay regions are continually evolving they have not, as yet, shown any clear tendency to move in any particular direction (though the Biscay

eddies illustrated in Fig.3(b) appear to have resulted from a S.E.
flow of water over abyssal depths ($\sim$ 4000 m) guided alongside the
Armorican slope).

An off-shelf southward flow from the Celtic Sea region can
occasionally be inferred in the winter period when cooler water
can be observed to extend S.E. across the shelf and into the
Biscay. The evidence so far from drifting buoys (indicating
surface water movement) in the N.W. Biscay also appears to
suggest a S.E. flow from about 49°N to 46°N (Fig. 2). This is
in marked contrast with the deeper current measurements which
show a well-established deep water (2000 m) flow ($\sim$ 2 cm s^{-1}
Fig. 2) N.W. along the slopes (Swallow et al., 1977; Dickson,
1982). To firmly establish the surface flows there is clearly a
need for more near-surface measurements covering different
seasons and spanning sufficiently long intervals of time so that
the wind conditions experienced are fairly representative of
the prevailing westerly winds for this region.

Warmer plumes have also occasionally been observed in the
winter period extending from the Spanish coast along the slopes
to 47°N and also along the central region of the Armorican shelf
($\sim$ < 100 m contour) in late autumn indicating an occasional N.W.
flow on the shelf as far north as the entrance to the English
Channel (Le Cann, 1982). At this time of year the off-shore
shelf water is generally cool and appears to be moving S.E. along
the shelf.

Whereas in the deeper waters some indication of water movement
is provided by the winter temperature structure, in the shallower
waters on the shelf the summer imagery has been more revealing.
Whilst the situation is clearly variable the tidal fronts have
shown a general tendency to be deformed from their predicted
positions thereby indicating residual currents. The Scilly Isles
plume tends to point northward and the Celtic Sea front shows a
marked meander. This suggests a northward movement of water from
the Scilly Isles toward the Bristol Channel and into the Irish
Sea on the eastern side of the St Georges Channel with a return
flow of cooler mixed water on the western side of the St Georges
Channel.

Current meter moorings were placed along the Irish coast to
see whether the flow indicated by the satellite imagery continued
S.W. as a coastal current. Although the residuals were marked

near the southwestern corner of Ireland indicating a coastal current, the flows further north were small ($\sim$ 1 cm s^{-1}) (Fig. 2). Cool water also appears occasionally along the Irish coast in this region, often extending in plumes from the headlands. This is thought to result from upwelling caused by the wind rather than an extension of cold mixed water from the Irish Sea. Cooler water stretching more than 100 km southwards from the southern meander of the Celtic Sea front can also be observed in late summer as the thermocline erodes and further studies are required to see whether this represents a significant compensation flow for the water entering the Irish Sea on the eastern side of the St Georges Channel.

4.2.   Turbidity

In the mixed waters of the English Channel the higher resolution C.Z.C.S. imagery has given some useful indications of water movement though many of the same features can be seen in the infra-red imagery. The sharpness of the contrast between the coastal and offshore water in the central English Channel is marked and suggestive of an advective flow (Fig. 15). A drifting buoy with a drogue at 30 m was released mid-channel in the narrows between the Isle of Wight and the Cotentin Peninsula three days after this image (8.3.83). It took $\sim$ 30 days to pass through the Dover Strait giving a flow of about 7 cm s^{-1}, somewhat larger ($\sim$ x3) than estimates for the mean transport through the Strait of Dover. A further surface drifting buoy placed in the coastal zone of Lyme Bay (in June 1983) (see Fig. 15) gave a flow approximately westwards from Lyme Bay and into Eddystone Bay with a residual current estimated as 5 cm s^{-1} over a period of 28 days. It remains an interesting subject of future research to establish whether or not mid-channel turbidity features can be correlated with the transport or the shear of waters in the English Channel.

Measurements showing that the C.Z.C.S. imagery provides a useful indication of the turbidity of the waters in the English Channel was obtained on a cruise of the R.V. Frederick Russell in March, 1983 and this can be appreciated by comparing Fig. 15 with the measurements of extinction coefficient made at sea with a transmissometer (see Fig. 16). No reason has yet been given as

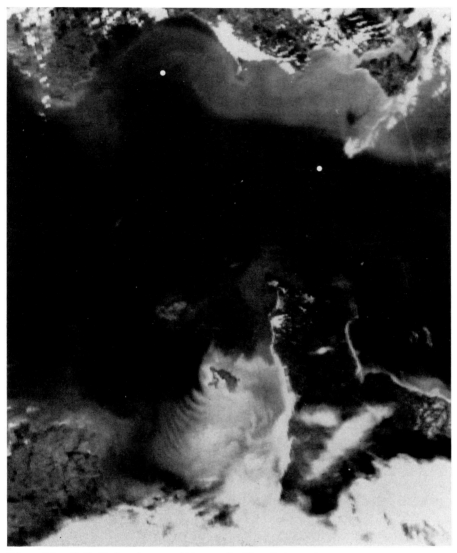

Fig. 15. C.Z.C.S. image (Channel 3) of 5.3.83 (1043 G.M.T.)
showing variations of turbidity between the central and coastal
waters of the English Channel (compare with Fig. 16) and sediment
plumes off headlands.   White dots mark the release positions of
surface drifting buoys (see text).

to why the clearer water is displaced to the French side of the
English Channel though it ought to be significant in determining
the development of the spring bloom in the Eastern English
Channel (Holligan et al., 1976).   Similar turbidity maxima and
minima have been observed in the Southern Bight of the North Sea

(Fig. 16) and their possible origins are discussed by Lee and Folkard (1969) and their relationship with diatom blooms is considered by Reid et al. (1983).

The C.Z.C.S. imagery also shows sediment plumes off headlands which also give a broad indication to the local net transport of water in some areas (see for example Fig. 15 and Fig. 2). In Fig. 15 an interesting pattern of fringes occurs to the S.W. of the Island of Jersey. These features actually occur in the sea and are not cloud or lee waves behind the island and were observed on the March 1983 cruise. These turbidity bands can also occasionally be observed in the infra-red satellite images and show up as adjacent warm and cold fringes of water. They tend to be more conspicuous in the winter months and most apparent just after spring tides suggesting a tidal origin. Clearly such an interesting phenomena requires further investigation at sea.

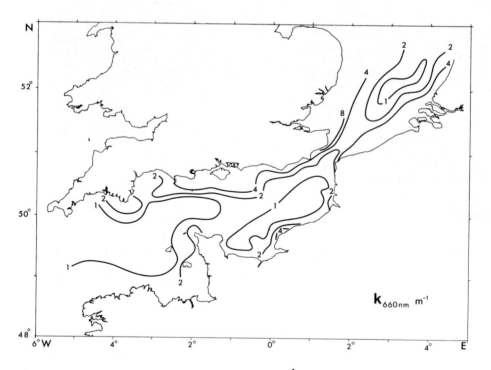

Fig. 16. The extinction coefficient (m^{-1}) for red light (660 nm) in the English Channel and Southern Bight of the North Sea (March 1983).

One possible explanation for their formation is that they are "tidal fringes" resulting from a residual flow (probably partly tidally induced) and the oscillatory tidal flow. It is assumed that each tide, one fringe of clear (salty, warm (winter) or cold (summer)) and one fringe of turbid (fresh, cold (winter) or warm (wummer)) water is introduced on the S.W. side of Jersey from a position N.E. of the island near the Paternosters. These are then stretched and diffused as they advect S.W. and as many as 28 tidal fringes ($\sim$ 14 tidal cycles) have been observed on a single image. Based on the separation, L, of the tidal fringes, the mean flow v, near the island is given as

$$V = \frac{L}{T} \sim \frac{3600}{44,700} \sim 0.08 \text{ ms}^{-1} \tag{6}$$

where T is the M_2 tidal period.

To see whether such a hypothesis could be valid it is instructive to estimate the lifetime of such features under the effects of continuing decay due to shear diffusion. Assume that in the direction of the residual flow the amplitude of the fringe signal has a sinusoidal character, with wave number $k = 2\pi/L$, then the timescale, τ, for a fringe to decay to 1/e of its initial expression following the direction of residual flow is

$$\tau \sim \frac{1}{k^2 K} \tag{7}$$

This neglects any effects due to density differences and assumes the fringes are much longer than their wavelength. Assume K is a constant shear diffusion coefficient of the form

$$K = \alpha u h \tag{8}$$

where α is a dimensionless coefficient, u is the vertically integrated maximum tidal current in this direction and h is the water depth. Taking $\alpha \sim 0.2$ (see Nihoul, 1982), $u \sim 0.25 \text{ ms}^{-1}$, and $h \sim 35$ m gives $\tau \sim 2$ days or ~ 4 tidal periods, which is not inconsistent with the hypothesis and shows that these features could persist for several days. They would then tend to be most conspicuous a few days after spring tides as observed.

5.   SUMMARY

   The influence of remote sensing on reorganising and initiating
new research programmes in the Biscay, Celtic Sea and English
Channel has been immense since the sensors have revealed
(sometimes for the first time) processes at work in the sea.   By
contrast the actual quantitative information that has been derived
from satellite images of the sea surface so far has been rather
limited.   Whilst there are many examples of satellite images
showing, more clearly than ever before, the kinds of features that
can be studied, good time sequences of specific processes or
events are still relatively scarce.   In this paper some examples
of observed sea surface features using radar, infra-red and
visible bands have been used in a qualitative manner to illustrate
shelf-break cooling, shelf-break phytoplankton, Biscay eddies,
coastal upwelling, tidal fronts, internal waves and water
turbidity structures.   In some instances these have been supported
with measurements obtained from sea programmes.   On the semi-
quantitative side, the imagery has provided an estimate for the
wavelength and phase speed for the internal tide, a value for
the residual flow in the Channel Isles region and an estimate of
the shear diffusion coefficient in this region based upon the
life-time of kilometre scale structures.

ACKNOWLEDGEMENTS

   The infra-red satellite images were supplied by P.E. Baylis,
Dundee University.

REFERENCES

Alpers, W. and Salusti, E., 1983.   Scylla and Charybdis Observed
   From Space.   Journal of Geophysical Research, 88: 1800-1808.
Coombs, S.H., Lindley, J.A. and Fosh, C.A., 1983.   Vertical
   distribution of Larvae of Mackerel (Scomber scombrus) and
   microplankton, with some conclusions on feeding conditions and
   survey methods.   In: (U.N.E.S.C.O.), Expert Consultation to
   examine changes in abundance and species composition of neritic
   fish stocks.   I.O.C., No. 33 (in press).
Dickson, R.R. and Hughes, D.G., 1981.   Satellite evidence of
   mesoscale eddy activity over the Biscay abyssal plain.
   Oceanologica Acta, 4: 43-46.
Dickson, R.R., 1983.   Global summaries and intercomparisons: flow
   statistics from long-term current meter moorings.   In:
   (A.R. Robinson), Eddies in marine science, Springer, New York
   (in press).

Frouin, R., 1981. Contribution à l'étude de la température de surface de la mer par télédetection au moyen de l'expérience spatial HCMM. Thèse Docteur de Specialité. L'université des sciences et Techniques de Lille, 145 pp.

Haury, L.R., Wiebe, P.H., Orr, M.H. and Briscoe, M.G., 1983. Tidally generated high-frequency internal wave packets and their effects on plankton in Massachusetts Bay. Journal of Marine Research, 41: 65-112.

Holligan, P.M., Pingree, R.D., Pugh, P.R. and Mardell, G.T., 1978. The hydrography and plankton of the eastern English Channel in March 1976. Annales Biologiques, 33: 69-71.

Holligan, P.M., Viollier, M., Harbour, D.S., Camu, P. and Champagne-Philippe, M., 1983. Satellite and ship studies of coccolithophore production along a continental shelf edge. Nature, 304: 339-342.

Korteweg, D.J. and de Vries, G., 1895. On the change of form of long waves advancing in a rectangular canal, and on a new type of long stationary waves. Philosophical Magazine, 39: 422-443.

Le Cann, B., 1982. Evolution annuelle de la structure hydrologique du Plateau continental au sud de la Bretagne. Modelisation numerique. Thèse Docteur Ingénieur. L'Université de Bretagne Occidentale.

Le Groupe Tourbillon, 1983. The Tourbillon experiment: a study of a mesoscale eddy in the eastern North Atlantic. Deep Sea Research, 30: 475-511.

Lee, A.J. and Folkard, A.R., 1969. Factors affecting turbidity in the Southern North Sea. J. Cons. int. Explor. Mer, 32: 291-302.

Madelain, F. and Kerut, E.G., 1978. Evidence of mesoscale eddies in the northeast Atlantic from a drifting buoy experiment. Oceanologica Acta, 1: 159-168.

Mazé, R., 1983. Movements internes induits dans un golfe par le passage d'une dépression et par la maree. Application au Golfe de Gascogne. Thèse Docteur. L'Universite de Bretagne Occidentale, 320 pp.

Nihoul, J.C.J., 1982. Hydrodynamic models of shallow continental seas. Application to the North Sea. Riga, Neupré, 198 pp.

Pingree, R.D. and Griffiths, D.K., 1978. Tidal fronts on the Shelf Seas around the British Isles. Journal of Geophysical Research, 83: 4615-4622.

Pingree, R.D., Holligan, P.M. and Mardell, G.T., 1979. Phytoplankton growth and cyclonic eddies. Nature, 278: 245-247.

Pingree, R.D. and Mardell, G.T., 1981. Slope turbulence, internal waves and phytoplankton growth at the Celtic Sea shelf-break. Phil. Trans. R. Soc. Lond. A, 302: 663-682.

Pingree, R.D., Mardell, G.T., Holligan, P.M., Griffiths, D.K. and Smithers, J., 1982. Celtic Sea and Armorican current structure and the vertical distributions of temperature and chlorophyll. Continental Shelf Research, 1: 99-116.

Simpson, J.H. and Hunter, J.R., 1974. Fronts in the Irish Sea. Nature, 250: 404-406.

Swallow, J.C., Gould, W.J. and Saunders, P.M., 1977. Evidence for a poleward eastern boundary current in the North Atlantic Ocean. ICES C.M. 1977/C:32: 11 pp (unpublished document).

REMOTE SENSING OF CHLOROPHYLL IN THE RED SPECTRAL REGION

S. LIN[1], G.A. BORSTAD[2] and J.F.R. GOWER[3]

[1]2nd Institute of Oceanography, Hangchow, People's Republic of China

[2]G.A. Borstad Ltd., 10474 Resthaven Drive, Sidney, British Columbia, Canada V8L 3H7

[3]Institute of Ocean Sciences, P.O. Box 6000, Sidney, British Columbia, Canada V8L 4B2

ABSTRACT

Reflectance spectra of natural water bodies usually exhibit a distinct peak at longer wavelengths which is caused by solar stimulated in vivo fluorescence of the phytoplankton chlorophyll a pigments. In most situations this peak is centered near 685 nm and a simple 3 point measure of its height (Fluorescence Line Height or FLH) can be used to estimate the concentration of near surface chlorophyll a. However, at least where blooms of the ciliate Mesodinium rubrum are observed, the emission wavelength is shifted towards longer wavelengths by as much as 15 nm. In a data set of 56 reflectance spectra obtained in coastal British Columbia waters, we find at least three principal Gaussian shaped emissions (at 682 nm, 692 nm and 710 nm) contributing to the apparent peak near 685 nm. Spectra from visibly discoloured blooms of Mesodinium rubrum could best be modelled by assuming large emissions at 710 nm and 692 nm with very small emission at 682 nm. For these spectra the concentration of extractable chlorophyll a (all forms) could be accurately estimated from the amplitude of the 710 nm Gaussian, while for all other spectra the prediction is derived from either the FLH method or an equation using the amplitudes of 682 nm, 692 nm and 710 nm.

INTRODUCTION

The goal of measuring ocean color is to estimate the concentration of certain constituents of the water, principally phytoplankton and inorganic suspended solids. For the measurement of the phytoplankton pigment chlorophylls, the green/blue radiance ratio (G/B ratio) is normally used (Clark, 1981, Gordon et al., 1980). The coastal zone color scanner (CZCS), which was launched in October 1978 was specifically designed for ocean color measurement and is now being used to map the distribution of chlorophyll a from space using this green to blue radiance ratio technique (Smith and Baker,

1982; Gordon et al., 1983). An alternative method of remotely measuring sea surface chlorophyll concentration has been developed at the Institute of Ocean Sciences, Patricia Bay (Neville and Gower, 1977. Gower, 1980; Gower and Borstad, 1981; Borstad and Gower, in press). In this case the solar stimulated in vivo fluorescence of chlorophyll a at 685 nm is used to remotely detect and quantify phytoplankton. The G/B ratio and the Fluorescence Line Height (FLH) methods employ upwelling signals in different spectral regions, and have differing advantages. In general the G/B ratio algorithm works best in oligotrophic regions such as the open ocean, where chlorophyll concentrations are low and the upwelling blue radiance is high. In the coastal zone however, greater chlorophyll absorption results in very low blue signals and the G/B ratio becomes increasingly insensitive·at higher pigment concentrations. By contrast, the strength of the fluorescence emission generally increases with increasing chlorophyll concentration to 50 mg/m^3 or more, and gives a positive indication of the presence of photosynthetic organisms. While it is reasonable to expect the remote fluorescence measurements to suffer from the same variations in fluorescence per unit chlorophyll as seen with standard filter fluorometer, our experience indicates that the errors are less than a factor 2 from a mean relationship (Gower, 1980; Borstad et al., 1981) and comparable to those encountered by workers using the green/blue method. The green/blue reflectance ratio can be altered by strong absorption at shorter wavelengths by dissolved organic material and this may confound the measurement of chlorophyll concentration in some coastal areas. An important advantage of the fluorescence line height (FLH) method for at least airborne survey operations, is that the fluorescence line is near the long end of the visible band where atmospheric Rayleigh path radiance and sky reflection are much reduced.

In the FLH method, the surface chlorophyll a concentration is inferred from the height of the fluorescence emission peak seen near 685 nm on reflectance spectra. Figure 1 demonstrates a simple 3 point calculation of FLH above a linear baseline for one spectrum. Similar repetitive calculations of FLH for reflectance spectra serially acquired from low flying aircraft have allowed us to successfully map the distribution of phytoplankton chlorophyll a in several regions including the western Canadian continental shelf (Fig. 2) and the eastern Canadian Arctic archipelago (Fig. 3).

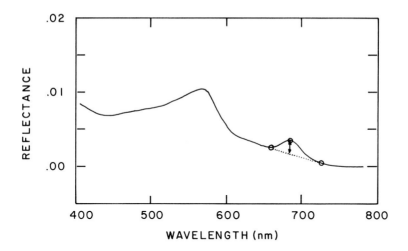

Fig. 1. A sample reflectance spectrum showing the calculation of Fluorescence Line Height (FLH) above a linear baseline between 643 nm and 729 nm.

METHODS

In order to more closely study the properties of the fluorescence line, we made a series of above-water measurements of the solar-stimulated *in vivo* fluorescence of chlorophyll *a* with in-water measurements of chlorophyll concentration and related parameters at 56 stations in southern Vancouver Island waters between July 25 and October 15, 1981 (Gower et al., 1983). Figure 4 shows all 56 reflectance spectra. These are reflectance factor spectra obtained from about 1.7 m above the water surface, viewing at the Brewster angle of incidence (53°) through a polarizer to reduce reflected sky radiance. Spectral resolution is about 12 nm. The highest reflectance observed is about 2%, with most spectra showing a maximum value near 1%. All spectra have been adjusted by a wavelength-independent reflectance contribution to compensate for residual glitter, foam or cloud reflection so as to give zero reflectance at 780 nm. This correction is usually small except where very large populations of the holotrich ciliate *Mesodinium rubrum* were present. The reflectance spectra from these visibly discoloured blooms were significantly different from the more normal spectra obtained for water containing smaller populations of commonly occurring diatoms and dinoflagellates. Spectra from the *Mesodinium* blooms showed a large anomalous peak at 710 nm, and arbitrarily

320

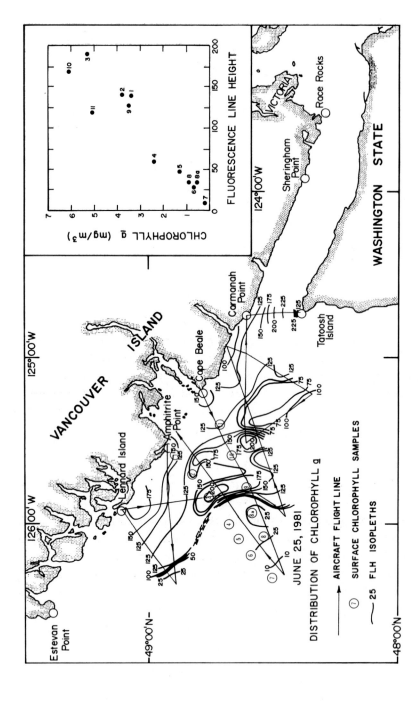

Fig. 2. The distribution of near-surface chlorophyll a off the west coast of British Columbia, Canada from airborne measurements in vivo fluorescence. The inset compares airborne FLH measurements with extracted chlorophyll a concentration in surface samples taken on the same day (from Borstad et al, 1981).

321

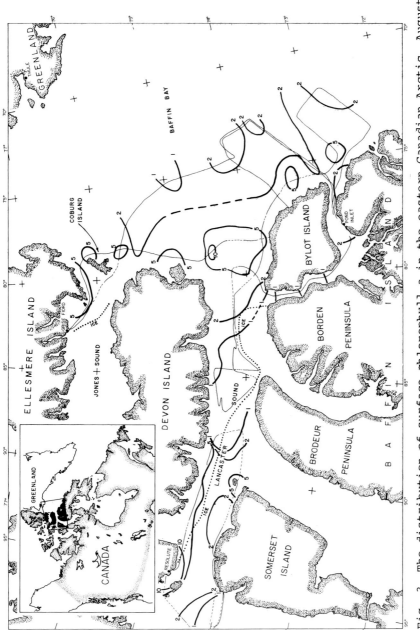

Fig. 3. The distribution of surface chlorophyll a in the eastern Canadian Arctic, August 18-27, 1979, based on airborne measurements of Fluorescence Line Height (FLH) (from Borstad and Gower, in press).

setting the reflectance at 780 nm to zero results in negative values
at shorter wavelengths. This does not affect our analyses. In
reflectance spectra from the <u>Mesodinium</u> stations, the fluorescence
emission normally found near 685 nm was located at wavelengths as
long as 712 nm, thus invalidating our usual FLH calculation which
is centered around 685 nm (Fig. 5).

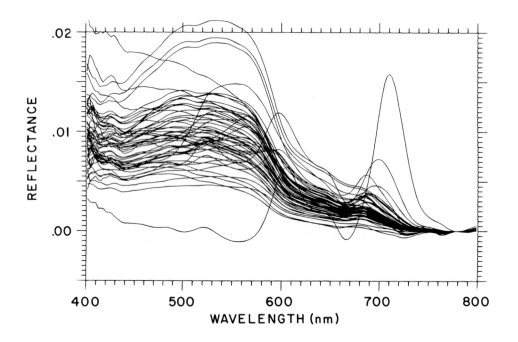

Fig. 4. Fifty-six reflectance factor spectra obtained in coastal
British Columbia waters between July 25 and October 15, 1981.

EIGENVECTOR ANALYSIS

In an effort to improve upon our 3 point FLH calculation and
also to study the information content of the red and blue ends of
the spectrum, we analyzed all 56 reflectance spectra again using
eigenvector analysis in different spectral regions (Gower et al.,
1983). Our results show that the variability of water radiance
spectra can be well described in terms of at most four vectors, in
agreement with the findings of Mueller (1973) and Doerffer (1981).
The variations found by the eigenvector analysis justify, to some
extent, our previous use of simpler parameters derived from the

spectra for estimating chlorophyll a concentration.

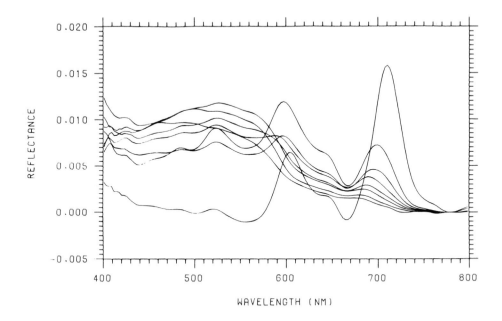

Fig. 5. Representative reflectance spectra for stations visibly discoloured by the ciliate <u>Mesodinium</u> <u>rubrum</u>.

Figure 6 illustrates the mean and characteristic spectral eigen-vectors in units of percent reflectance for all spectra. Figure 7 demonstrates a good relationship (R = .945) between measured chlorophyll concentration and that calculated using 4 eigenvalues. This relationship is also valid for the <u>Mesodinium</u> spectra and chlorophyll a concentrations as high as 35 mg/m^3.

For different spectral regions, our analysis shows:

(1) The variability in eigenvectors is due to the combination of absorption by chlorophyll and phaeopigments as well as scattering by the associated cellular material in phytoplankton and other sus-pended matter.

(2) We obtain as good a correlation with chlorophyll a using data from the red spectral region as from the blue/green region, showing that for our observations, as much information is present at the longer wavelengths, in spite of the lower signal levels there.

Examination of the reflectance spectra and eigenvalues from the

<u>Mesodinium</u> spectra have also led us to attempt to separate indepen-
dent absorption and fluorescence peaks in the red spectral region.

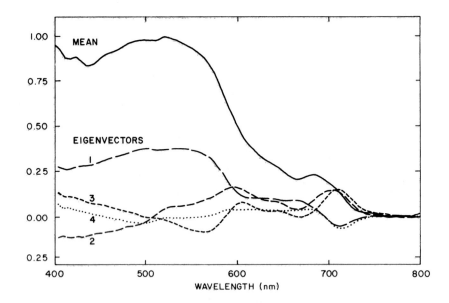

Fig. 6. The mean spectrum and characteristic spectra (eigenvectors)
in units of percent reflectance for all spectra.

ABSORPTION AND FLUORESCENCE IN THE RED SPECTRAL REGION

For this exercise, all data were analysed again using eigenvector
analysis after a linear sloping baseline (from 643 to 780 nm) was
subtracted from the diffuse reflectance spectra. Subtraction of a
linear baseline is somewhat arbitrary but a more realistic choice
depends not only upon the interaction of water absorption and
particle backscatter, but also the fluorescence and absorption of
other pigments in the range 600 to 750 nm. The difference between
the linear baseline and a more correct (curved) baseline will
contribute to our vectors, and this must be kept in mind when interpreting this data.

Five new eigenvectors were obtained, which together can explain
99.96% of the total variance of the sample (Fig. 8, upper panels).
The first three vectors contain several peaks and dips, presumably
representing absorption and emission. Using a Gaussian curve
fitting method to fit these peaks, we obtained five separate
Gaussian curves (Fig. 8, bottom panel). The center wavelengths and
half-power band widths are listed in Table 1. As shown in this

Table, we obtain close agreement with biochemical and biophysical
laboratory studies for all but the 667 nm Gaussian. Why this should
be so is not clear at present, but is under investigation.

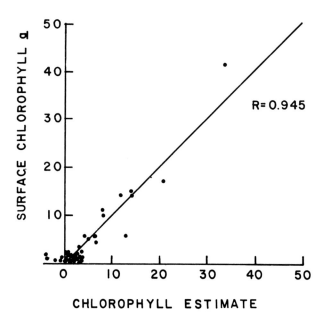

Fig. 7. The relation between chlorophyll a concentration calculated
by eigenvector analysis and the actual extracted chlorophyll a
concentration (both mg/m³).

FLUORESCENCE LINE SHIFT AND MESODINIUM FLUORESCENCE

Now, if the five independent Gaussian peaks are used as new vector components, all 56 reflectance spectra can be reconstructed by
adding an appropriate fraction of each to the sloping baseline.
For all stations, the amplitudes (= fractions) for the peaks at 667
nm and 735 nm are negative. This is partly because of the choice
of a linear baseline which is too high in these regions, however,
while the reflectance spectra (Fig. 4) suggest a positive signal near
735 nm, there is a strong dip near 667 nm. We are interpreting this
signal as an absorption feature, while all of the others represent
in vivo fluorescence emitted upon solar stimulation. This agrees
with laboratory measurements (Goedheer, 1972; Govindjee et al.,
1979; Prézelin, 1981). Figures 9 and 10 demonstrate the reconstruction of an observed reflectance spectrum for two stations representing a 'normal' phytoplankton population 90% dominated by diatoms
(Fig. 9) and a visually discoloured bloom of Mesodinium rubrum

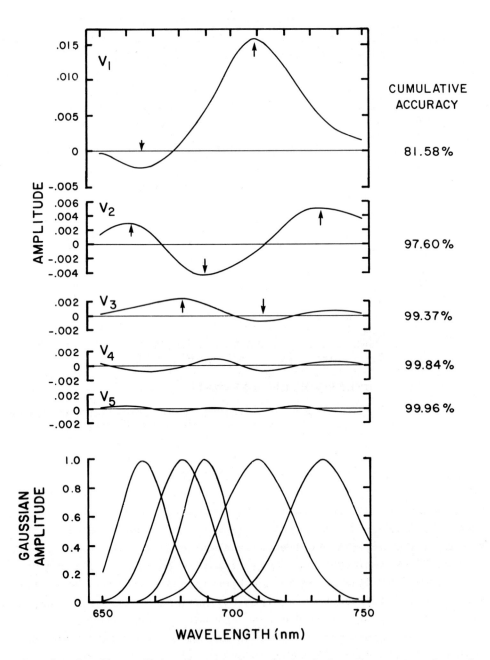

Fig. 8. The first five eigenvectors derived for the spectral region 650 to 750 nm, after removal of a linear baseline. The five Gaussian emissions (below) are derived from the first three vectors in upper panels through a curve fitting routine.

TABLE 1

Comparison of the center wavelengths of the Gaussian shaped signals
derived from eigenvectors of field reflectance spectra, with data
from laboratory measurements.

Gaussians derived from reflectance spectra		Laboratory measurements		
Half-power band width	Center wavelength	Center wavelength	Probable source	Reference
22 nm	667 nm	662-683 nm	absorption	Govindjee and Braun, 1974.
25 nm	682 nm	681-688 nm	F_{685} in PSII*	Brown, 1967; Govindjee et al., 1979.
21 nm	692 nm	690-695 nm	F_{690} in PSII*	Brown, 1967; Govindjee et al., 1979.
32 nm	710 nm	705-715 nm	F_{700} in PSI*	Brown, 1967; Govindjee et al., 1979.
32 nm	735 nm	730-740 nm	"vibrational satellite"	Brown, 1967; Govindjee et al., 1979.

*PSI and PSII (Photosystems I and II) are functional divisions of
the plant photosynthetic mechanism.

(Fig. 10). The upper panels show the observed reflectance spectra,
while the bottom panels illustrate the residual reflectance signal
not accounted for by the recalculation of the reflectance spectrum
from the Gaussian components. In both situations the error is small
relative to the observed spectrum.

Figures 9 and 10 also demonstrate the apparent fluorescence line
shift. In Figure 9, for a diatom population having a chlorophyll a
concentration less than 1 mg/m³, the 682 nm component is dominant
and the fluorescence line peak appears near 682 nm on the reflectance spectrum. Where Mesodinium rubrum blooms were observed (Fig.
10), the 710 nm Gaussian line is strongest and the apparent fluorescence line is shifted to longer wavelengths.

It is evident from this discussion that the apparent fluorescence
line can be described in terms of several Gaussian shaped emissions.
In order to explore the possibility of remotely obtaining information regarding the type of phytoplankton from the reflectance spectra, we next divided the 39 stations for which we had taxonomic
data into four groups according to their dominance by various phytoplankton (greater than 60% diatoms, dinoflagellates, unarmoured

328

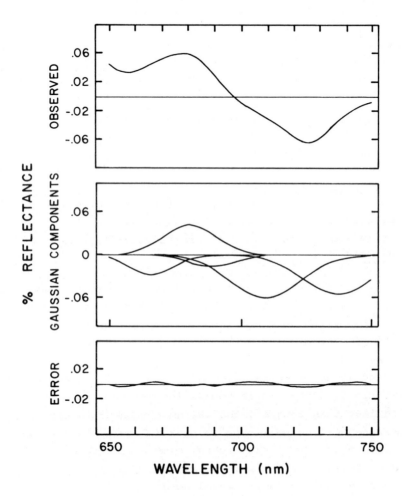

Fig. 9. Reconstruction of an observed reflectance spectrum (for a diatom population) through the use of Gaussian shaped components. The bottom panel shows the difference between the measured and reconstructed spectra.

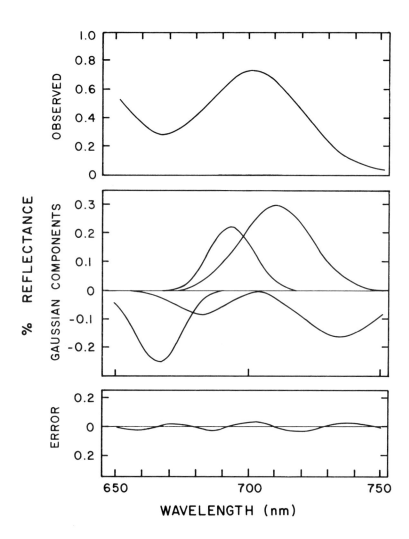

Fig. 10. Reconstruction of an observed reflectance spectrum (for a visually discoloured bloom of <u>Mesodinium</u> <u>rubrum</u>) through the use of Gaussian shaped components. The lower panel illustrates the difference between the measured and reconstructed spectra.

flagellates or Mesodinium rubrum.

We find that each different algal group has different optical properties. For example, even though the extracted chlorophyll a content is the same for two groups, the most significant fluorescence line may be located at different wavelengths. Table 2 illustrates this point. The chlorophyll a concentration was 6.0 mg/m^3 at both stations 10 and 34. However, for station 10, the main fluorescence line is at 682 nm, while for station 34, it is located at 692 nm.

TABLE 2

Comparison of Gaussian amplitudes for stations having similar extracted chlorophyll a concentration but different phytoplankton.

Station	Dominant Phyto-Plankton	Chlorophyll a concentration	A_{682} (x 10^3)	A_{692} (x 10^3)	A_{710} (x 10^3)
10	diatoms	6.0	1.26	0.33	-0.05
34	Mesodinium	6.0	0.23	0.94	0.49
3	flagellates	14.1	1.46	1.28	0.33
33	Mesodinium	14.0	0.19	1.01	0.54
2	dinoflagellates	2.1	0.43	-0.6	-0.51
38	Mesodinium	2.8	0.29	0.25	-0.60

Figure 11 illustrates the relationship between the amplitudes at 710 nm and 682 nm for the four groups. For all but the Mesodinium spectra the amplitude at 710 nm increases at about one half the rate of the 682 nm amplitude. Note that the artificial linear base line is responsible for the negative 710 nm amplitudes. We are not interpreting these as absorption and visual inspection and the literature supports this. Similarly, the choice of a linear baseline colours our interpretation of the apparent slope of the Mesodinium data in Figure 11. Inspection of the spectra from the visually discoloured stations (Fig. 3) shows that the short wavelengths end of the baseline is being lifted by phycoerythrin fluorescence near 600 nm. We conclude that Mesodinium rubrum has only a small and constant amount of chlorophyll a fluorescence at 682 nm (absolute concentrations are difficult to calculate at this

stage in our analysis). This is interesting from a biological
point of view because although <u>Mesodinium</u> <u>rubrum</u> is capable of
photosynthesis it is not a plant, but a protozoan containing what
are regarded as "incomplete symbionts" - essentially just chloro-
plasts (Taylor, Blackbourn and Blackbourn, 1978). The absence of
fluorescence at 682 nm presumably means that this organism lacks the
form of chlorophyll <u>a</u> which normally constitutes 80% of the cell
total in other kinds of phytoplankton (Prézelin, 1981). The many
forms of chlorophyll <u>a</u> (which are not differentiated in the routine
extractive procedures used by oceanographers) are unequally distributed within the plant photosynthetic mechanism (Govindjee and Braun,
1974) and there are some indications that the relative amounts of
fluorescence changes between algal types (Goedheer, 1972) or in
ageing or lowlight adapted cells (Brown, 1967).

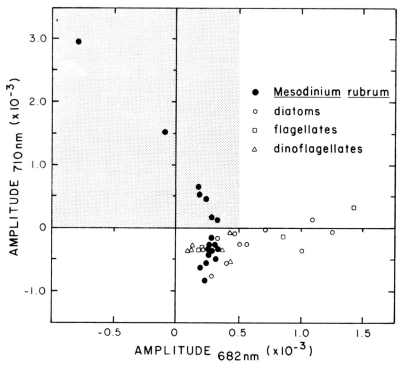

Fig. 11. The relationship between the Gaussian amplitudes at 682 nm
and 712 nm in 39 reflectance spectra from British Columbia coastal
waters. Spectra from <u>Mesodinium</u> <u>rubrum</u> populations show a significantly different signature, thus allowing this species to be
remotely differentiated from other forms of phytoplankton within
the stippled zone (extracted chlorophyll <u>a</u> concentration 4 - 199
mg/m^3).

ESTIMATION OF CHLOROPHYLL A CONCENTRATION FOR DIFFERENT PHYTOPLANK-
TON POPULATIONS

We are currently investigating whether it is possible to use the
methods described here to remotely detect either gross taxonomic or
physiological changes in a phytoplankton population. For this data
set we can differentiate Mesodinium rubrum populations from those
of dinoflagellates, diatoms and flagellates on the basis of the
710 nm emission. Where the amplitude of the 710 nm Gaussian is negative or the 682 nm Gaussian is greater than 0.0005 (unstippled area
in Fig. 11) we can calculate the chlorophyll concentration according
to the formula:

$$mgChl\ a/m^3 = 0.4 + (28.34A_{682} + 77.66A_{692} - 16.48A_{710}) \times 10^2 \quad (1)$$

Figure 12 illustrates the agreement between this calculation and
extracted chlorophyll a concentration for populations dominated by
dinoflagellates, diatoms or flagellates. The correlation coefficient is 0.96 while the scatter about the 1:1 line is about ± 1.0
mg/m^3. This is better than using the eigenvector analysis (Fig. 7).

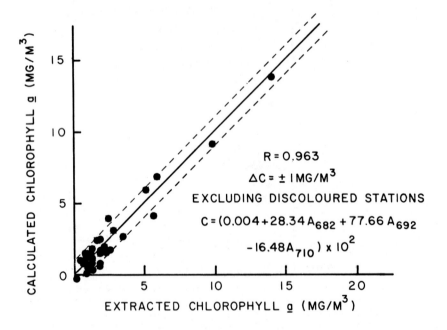

Fig. 12. The agreement between the extracted chlorophyll a concentration and that calculated on the basis of three Gaussian signals
at 682 nm, 692 nm and 710 nm.

Where the amplitude of the 710 nm Gaussian is positive and the 682 nm Gaussian less than 0.0005 (stippled area in Fig. 11), we can remotely classify the dominant organism as <u>Mesodinium</u> and calculate the concentration of chlorophyll <u>a</u> from the 710 nm amplitude.

$$mgChl \underline{a}/m^3 = 5.19 + (5.65 A_{710}) \times 10^2 \qquad (2)$$

Figure 13 illustrates the agreement between this calculation and extracted chlorophyll over the range 4 to 199 mg chlorophyll a/m^3. The correlation coefficient in this case is 0.95, while the scatter about the line is as great as $\pm$ 4 mg/m^3.

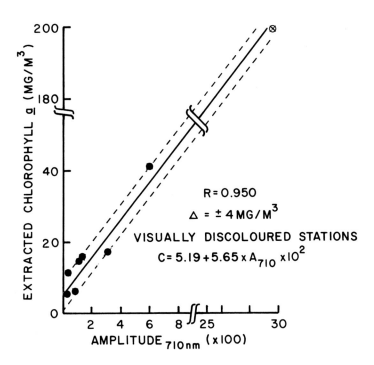

Fig. 13. The relationship between the amplitude of the 710 nm Gaussian emission and the extracted chlorophyll <u>a</u> concentration for seven spectra obtained from visually discoloured blooms of <u>Meso-</u><u>dinium</u> <u>rubrum</u>. Regression and correlation coefficients do <u>not</u> include highest point.

SUMMARY AND CONCLUSIONS

The fluorescence line height (FLH) method can be successfully employed to remotely measure the chlorophyll a concentration in many oceanic areas, however, where large blooms of the ciliate Mesodinium rubrum are encountered, the accuracy of the FLH calculation is significantly affected by an apparent shift in the emission wavelength.

From an analysis of 56 reflectance spectra obtained in coastal British Columbia waters, we find at least three principal Gaussian shaped fluorescence lines, located at 682 nm, 692 nm and 710 nm. In this data set spectra from visibly discoloured blooms of Mesodinium rubrum could best be modelled by assuming large emissions at 710 nm and 692 nm, with very small emission at 682 nm. The concentration of extractable chlorophyll a (all forms) for these populations could be accurately estimated from the height of the 710 nm Gaussian. For all other spectra, where the apparent fluorescence line is located near 685 nm, the extractable chlorophyll a is estimated either by the FLH method or an equation employing the amplitudes at 682 nm, 692 nm and 710 nm.

REFERENCES

Borstad, G.A., Brown, R.M., and Gower, J.F.R., 1981. Airborne
    remote sensing of sea surface chlorophyll and temperature along
    the outer British Columbia coast. Proceedings of the 6th
    Canadian Symposium on Remote Sensing, Halifax, N.S., pp. 541-547.
Borstad, G.A. and Gower, J.F.R., 1983. A ship and aircraft survey
    of phytoplankton chlorophyll distribution in the eastern
    Canadian Arctic. Arctic (in press).
Brown, J.S., 1967. Fluorometric evidence for the participation of
    chlorophyll a - 695 in systems of photosynthesis. Biochem.
    Biophys. Acta., 143:391-398.
Clark, D.K., 1981. Phytoplankton algorithms for the Nimbus-7 CZCS.
    In: J.F.R. Gower (Editor), Oceanography From Space, Plenum Press,
    New York, pp. 227-228.
Doerffer, R., 1981. Factor analysis in ocean colour interpretation.
    In: J.F.R. Gower (Editor), Oceanography From Space, Plenum Press,
    New York, pp. 339-345.
Goedheer, J.C., 1972. Fluorescence in relation to photosynthesis.
    Ann. Rev. Plant Physiol., 23:87-112.
Gordon, H.R., Clark, D.K., Brown, J.W., Brown, O.B., Evans, R.H.,
    Broenkow, W.W., 1983. Phytoplankton pigment concentrations in
    the Middle Atlantic Bight: a comparison of ship determinations
    and CZCS estimates. Appl. Optics, 22:20-36.
Gordon, H.R., Clark, D.K., Mueller, J.L. and Hovis, W.A., 1980.
    Phytoplankton pigments from the Nimbus-7 Coastal Zone Colour
    Scanner: Comparisons with surface measurements. Science,
    210:63-66.
Gower, J.F.R., 1980. Observations of in situ fluorescence of
    chlorophyll a in Saanich Inlet. Boundary Layer Meteorology,
    18:235-245.
Gower, J.F.R. and Borstad, G.A., 1981. Use of in vivo fluorescence
    line at 685 nm for remote sensing surveys of surface chlorophyll
    a. In: J.F.R. Gower (Editor), Oceanography From Space, Plenum
    Press, New York, pp. 329-338.
Gower, J.F.R., Lin, S., and Borstad, G.A., 1983. The information con-
    tent of different optical spectral ranges for remote chlorophyll
    estimation in coastal waters. Int. J. Remote Sensing (in press).
Govindjee and Braun, B.Z., 1974. Light absorption, emission and
    photosynthesis. In: W.D.P. Stewart (Editor), Algal Physiology
    and Biochemistry, Univ. California Press, Berkeley, pp. 346-390.
Govindjee, Wong, D., Prezelin, B.B. and Sweeney, B.M., 1979.
    Chlorophyll a fluorescence of Gonylaulax polyedra grown on a
    light - dark cycle after transfer to constant light. Photochem.
    Photobiol., 30:405-411.
Mueller, J.L., 1973. The influence of phytoplankton on ocean
    colour spectra. PhD. Thesis, Oregon State University, Corval-
    lis.
Neville, R.A. and Gower, J.F.R., 1977. Passive remote sensing of
    phytoplankton via chlorophyll a fluorescence. J. Geophys. Res.,
    82:3487-3493.
Prezelin, B.B., 1981. Light reactions in photosynthesis. In: T.
    Platt (Editor), Physiological bases of phytoplankton ecology.
    Can. Bull. Fish. Aquat. Sci., 210:1-42.
Smith, R.C. and Baker, K.S., 1982. Oceanic chlorophyll concentra-
    tions as determined by satellite (Nimbus-7 Coastal Zone Colour
    Scanner). Mar. Biol., 66:269-280.

Taylor, J.F.R., Blackbourn, D.J. and Blackbourn, J., 1971. The red-water ciliate _Mesodinium_ _rubrum_ and its "incomplete symbionts": a review including new ultrastructural observations. J. Fish. Res. Bd. Canada, 28:391-407.

SATELLITE REPRESENTATION OF FEATURES OF OCEAN CIRCULATION
INDICATED BY CZCS COLORIMETRY

C.S. YENTSCH

Bigelow Laboratory for Ocean Sciences, West Boothbay Harbor,

Main 04575, U.S.A.

ABSTRACT

   Coastal Zone Color Scanner (CZCS) images have been used to
demonstrate that the major factors which influence the patterns of
ocean color and hence the abundance of phytoplankton are associated
with the density discontinuities of large scale ocean currents.
This argues that variations in color in large scale patterns are
reflecting phytoplankton growth. Pigment patterns, therefore, are
not passive tracers of surface water movement.

INTRODUCTION

   There is now a considerable number of CZCS images which allow the

biological oceanographer to visually see patterns of phytoplankton

pigments over large regions of the earth's oceans. In examining

these images, one's first impression is that the ocean is characte-

rized by highly diverse patterns of pigment concentrations. It is

also evident that the spatial magnitude of these patterns differ.

Immediately we can ask : "Are these patterns the result of spatial

movements of phytoplankton ?" "Can phytoplankton be considered a

conservative tracer of the water masses, thereby producing patterns

similarly seen by adding cream to a teacup ?" Or, "Are these

patterns explained in terms of factors other than horizontal transport,

specifically those factors which we believe regulate the growth

and abundance of phytoplankton in the oceans ?" These questions are

important to oceanographers since the distribution of phytoplankton

in time and space, and the mechanisms controlling this distribution,

have been obtained largely by one-dimensional shipboard observa-

tions which are limited in coverage of both time and space : the

need for remote sensing is driven by the desire to view the

enormity of ocean space in synoptic fashion and to test wether or
not we have not biased our impressions by quasi-synoptic observation on ships.

This paper has two main goals. First, in a general sense, to
acquaint the uninitiated reader with some of the factors affecting
large scale distribution of phytoplankton in the oceans and to
demonstrate how the spatial distributions are viewed from space.
The second goal of the paper is to demonstrate and interpret the
large scale patterns of phytoplankton in terms of the major planetary inertial forces that are operating on water masses. I will
argue that the spatial patterns are reflecting the degree of
buoyant forces in the water mass. That is, spatial changes that
one observes in these images are regulated by the intensity of
vertical mixing throughout the water column. If correct, then the
large scale patterns, and perhaps the small scale patterns as well,
are reflecting the net growth of phytoplankton. In other words,
the distribution of phytoplankton pigment abundance observed in the
surface waters of the oceans is representing growth processes and
not merely the redistribution of abundance.

To give substance to these goals, I will utilize satellite
images and conceptual models as well as water column observations.
I have specifically chosen regions of the oceans where fluid forces
favor the destruction of buoyancy in the water column, thereby
promoting vertical mixing. These forces are derived from the shear
effects associated with major frontal regions of ocean currents,
the rotary motions of mesoscale ocean eddies, and friction from
the interaction of tidal flow across shallow waters.

Large scale features of phytoplankton distribution associated with general circulation.

For phytoplankton, the extremes of poverty and luxury are defined
by the oligotrophic central gyres of the ocean on one hand, and
the eutrophic waters that lie adjacent to major continents on the
other hand. Between these extremes are sharp gradients of phytoplankton abundance which are correlated with water masses which
have an extreme baroclinicity and intense horizontal advection.

The effects of ocean currents were seen from space first in
satellite thermal images. However, more recently, CZCS colorimetry
has demonstrated the sharp color discontinuities associated with
ocean currents, thus delineating marked gradients in phytoplankton
abundance. The question is : How does large scale flow change the

distribution of phytoplankton ?

The density field of large ocean currents are "baroclinic analogues" of upwelling and represent the largest, perhaps most important, mechanism in the world's ocean of vertical transport of nutrients (Yentsch, 1974). Some of the best examples of these extensive color fronts representing discontinuities in phytoplankton abundance are found in regions occupied by the western boundary current systems. Figure 1 is CZCS Orbit 02646 that features the Gulf Stream system from Cape Hatteras to Yarmouth, Nova Scotia. The reader's attention is called to the delineation by color of slope and Gulf Stream waters.

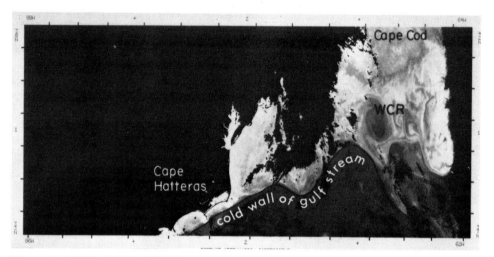

Fig. 1. CZCS Orbit 02646 featuring the Gulf Stream system from Cape Hatteras to Yarmouth, Nova Scotia.

The process which is responsible for the delineation of color concerns augmentation of phytoplankton growth which is directly associated with the geostrophic flow. The first fluid dynamic model of the Gulf Stream was produced by Carl Rossby. He considered the Stream a major jet driving into a non-rotating stratified fluid (Fig. 2). When the earth's rotation (C_f) was considered as a balance to the pressure gradient (P_g), the Rossby model predicted that secondary cross-stream flow would be associated with the horizontal advection. This cross-stream flow was transported along lines of equal density from the Sargasso Sea into the slope and coastal waters off New England. The important aspect of this model of phytoplankton growth is that it demonstrates that by isopycnal transport, nutrients necessary for growth traverse great distances

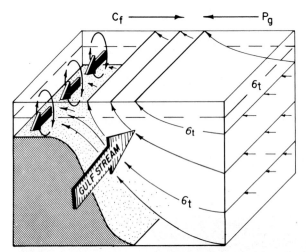

Fig. 2. The Rossby (1936) model of the Gulf Stream system off New England.

horizontally and vertically - that is, from the deep waters of the Sargasso Sea to the surface waters of the euphotic zone in slope waters off New England. Examples of the effect of this transport can be seen by comparing density structure across the Gulf Stream (Fig. 3) with the distribution of a limiting nutrient such as nitrate-nitrogen in the same section (Fig. 4). Facing into the picture, one sees line of equal density intersecting at station 9, which is referred to as the "cold wall", since cooler, deeper waters are elevated to that side of the Gulf Stream. The enrichment process is signaled by the fact that lines of equal density are mirrored by lines of equal distributions of nitrate which, as mentioned above, is the limiting nutrient for phytoplankton growth in these waters. It should be noted that the fluid dynamics behind movements of the water along the isopycnals, is still not well understood and the resultant magnitude of vertical transport is not well known. In general, the cause of the movement along isopycnals can be considered an imbalance between the pressure gradient (P_g) and the Coriolis forces (C_f) associated with the mass transport of the Gulf Stream itself. Regardless of the cause, the fertility of the waters lying adjacent to the main thrust of the Gulf Stream (in the cold wall) can be traced along lines of equal density from the north central Sargasso Sea to the cold wall of the Gulf Stream (Fig. 5). The enriched water entering the euphotic zone in the cold wall causes a marked discontinuity in the spatial abundance of phytoplancton chlorophyll. It is this varia- tion that one clearly sees from space by way of CZCS colorimetry as a marked difference in water color which distinguishes the slope waters from that in the Gulf Stream.

Other examples of phytoplankton augmentation associated with major ocean currents can be observed in CZCS images of Florida and the western Gulf of Mexico. In this region, the thermal loop current forms a front which is due to the entry of equatorial water into the western Gulf of Mexico through the Straits of Yucatan (Fig. 6). The equatorial water penetrates as far north as

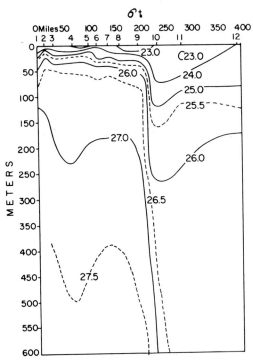

Fig. 3. Distribution of density (σt) across the Gulf Stream. Section between Cape Cod and Bermuda. Chain 37 July 1963. (Yentsch, 1974)

27° N and essentially encompasses most of the region of the western Gulf of Mexico. The CZCS colorimetric pattern of this image corre- lates with the general thermal pattern shown in the infrared image (Fig. 6). This correlation shows that warm equatorial waters are associated with low concentrations of phytoplankton pigment and

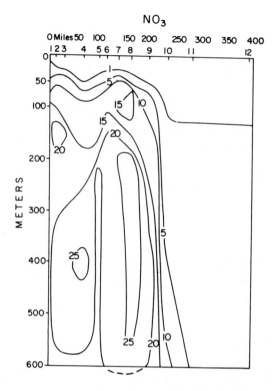

Fig. 4. Distribution of nitrate-nitrogen (μg-at/ℓ) across the Gulf Stream. Section between Cape Cod and Bermuda. Chain 37 July 1963. (Yentsch, 1974).

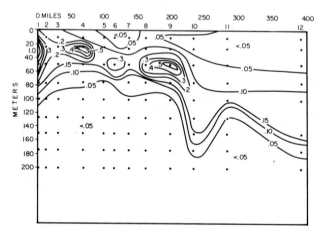

Fig. 5. Distribution of chlorophyll (μg-at/ℓ) across the Gulf Stream. Section between Cape Cod and Bermuda. Chain 37 July 1963. (Yentsch, 1974).

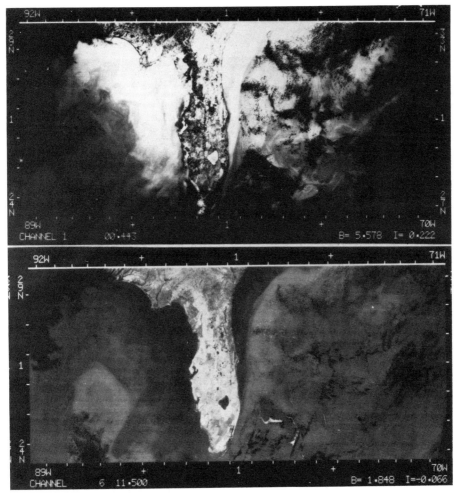

Fig. 6. NIMBUS-7 imagery from Orbit 1965 on 15 March 1979. CZCS Channel 1 (443 nm) image where light tone denotes high attenuation of blue due to phytoplankton chlorophyll. Channel 6 images of sea-surface temperature variation in which the dark tone depicts cold water.

cooler waters with high levels of phytoplankton pigment. The nutrient enrichment process and increased phytoplankton abundance are a combination of the flow of the loop current and the constraints placed on that current by the shape of the Florida peninsula continental platform. The position of the pigment fronts outlined from the image follow the trend of the isobaths along both coasts of the Florida peninsual (Fig. 7). The general position of these fronts is interpreted to be associated with the mass transport on

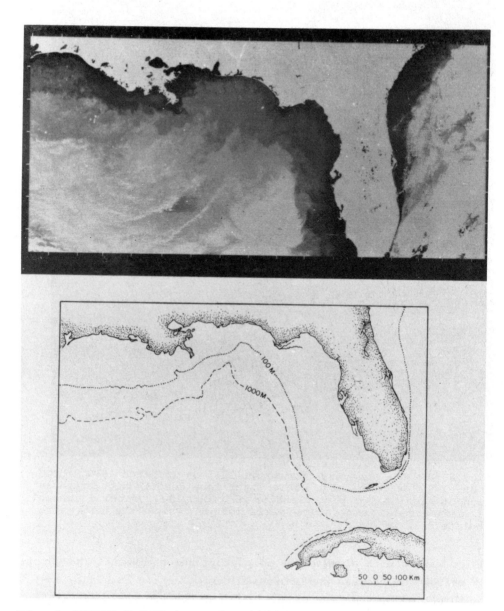

Fig. 7. NIMBUS-7 CZCS image from Orbit 30 on 2 November 1980, of the Florida region showing chlorophyll concentration (dark) on the coastal shelf (upper) and the major bathymetric features of the region (lower).

either side of the peninsula. This is confirmed by comparing the dynamic topography on the western side of the Gulf (Fig. 8). The distribution of sea level height shows that channel constraints of the mass flow by the Florida escarpment augments the horizontal velocity of the flow. Along with this augmentation of flow, an imbalance between the Coriolis and pressure forces create the isopycnal flow which causes the enrichment of the waters adjacent to the peninsula. Therefore, it is through these processes that

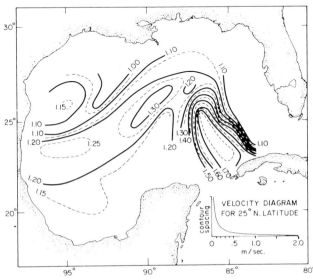

Fig. 8. Dynamic topography of the Gulf of Mexico region (Nowlin and McLellan, 1967).

we can account for the color outline of the general path of the current.

In summary, satellite imagery shows a correspondence between pigment-ocean color and major ocean currents. This demonstrates that the aegeostrophic forces associated with major ocean currents markedly influence phytoplankton growth and hence, their distribution. These large scale processes in effect, dictate the <u>major patterns of growth and abundance</u> of phytoplankton in the oceans.

<u>Mesoscale eddies associated with western boundary flow</u>

Mesoscale eddies are common features of the Gulf Stream system especially in the region north of Cape Hatteras. These eddies or Gulf Stream rings, as they are often called, form from extensive meanders of the Gulf Stream system (Fig. 9). Such meanders at

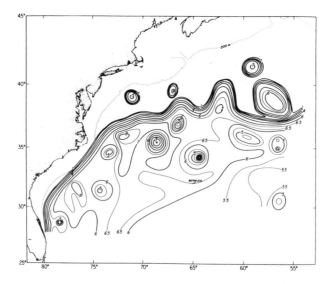

Fig. 9. Chart of the depth, in hundreds of meters, of the iso-
thermal surface, showing the Gulf Stream, nine cyclonic rings,
and three anticyclonic rings. Contours based on data obtained
between 16 March and 9 July 1975.(From Richardson et al., 1978).

times close (pinch off) portions of water masses on either side of
the Gulf Stream. The eddies formed by "pinching off" a warm core
of Sargasso Sea water are referred to as warm core rings and reside
in the slope water to the west of the Gulf Stream (Fig. 10). Cold
core rings are the reverse in that by the "pinching off" process,
slope water is entrained in the center. These cold core rings
generally move into the Sargasso Sea (Fig. 11). These rings were

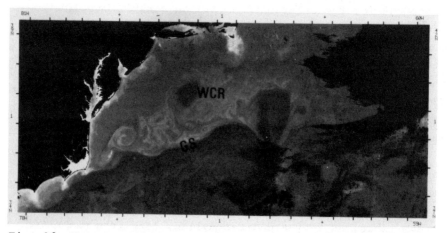

Fig. 10. Warm core ring (center) and new ring forming on right.

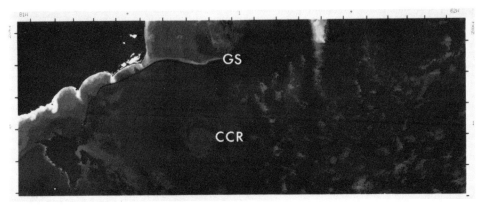

Fig. 11. Cold core ring (CCR) off Cape Hatteras.

first observed by Fritz Fuglister using shipboard temperature
measurements, however, both warm and cold core rings, because of
their sharp thermal gradients, are easily identified in satellite
thermal imagery. Satellite observation by CZCS colorimetry has
demonstrated that both warm core and cold core rings are also well
defined in terms of their differences in color : The sharp thermal
gradient as seen by the satellite, are mirrored by gradients in
phytoplankton pigment (Gordon et al., 1982). The question we can
now ask is : Why is this so ?

The rotary motion of ocean eddies to phytoplankton growth
concerns changes in the vertical distribution of the density field
within the eddy. If we assume that phytoplankton growth is nutrient limited and distribution of nutrients is reflected by the
density field, then the following concepts (Fig. 12) influence
spatial patterns of growth throughout the eddy. N^+ and N^- represent
two water masses of nutrient-rich, cold, dense and nutrient-poor,
warm buoyant water, respectively. These are enclosed in a cylinder
which simulates the dimensions of an oceanic eddy. The two water
masses are separated by the density nutrient boundary layer (N_b)
which for this discussion we can refer to as the thermocline. In
the non-rotational stationary mode, the boundary between the two
water masses is horizontal across the cylinder. However, when the
cylinder is rotated with velocities in the surface being somewhat
greater than at depth, the Coriolis and other inertial forces will
be balanced by the pressure gradient created by the geostrophic
flow within the eddy. In the anti-cyclonic mode, sea surface level
domes up around the axis (warm core) while in the cyclonic mode,

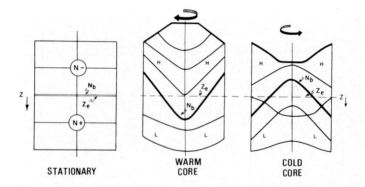

Fig. 12. Geostrophic relationships in warm and cold core rings. N_b, nutrient boundary; Z_e, euphotic zone; H and L are the high and low velocities.

it will be depressed in the axis (cold core). In the anti-cyclonic eddies, such as the warm core ring, the lighter water will accumulate at the center and the heavy water will be swept to the rim of the eddy. Assuming that the volume of the eddy is being maintained, the boundary dips downward towards the axis and upward towards the rim of the eddy. The reverse situation occurs in the cold core ring. If the eddy is illuminated from the surface, and the photic layer (Z_e) resides at a comparable depth and boundary layer, we can see why productivity is enhanced due to the upward displacement of nutrient rich water. This upward displacement of the nutrient boundary layer allows vertical mixing to easily transport nutrients to the euphotic zone. Therefore, the spatial pattern of phytoplankton distribution reflects relative nutrient addition to the euphotic layer by the differences vertically in the level of the boundary between the two water masses.

The explanation for the observed distribution of phytoplankton pigments in rings argues that geostrophic principles apply to these rings. Implicit to this nutrient enrichment hypothesis is the idea that the rotary motion induces nutrient transport along isopycnals and phytoplankton production occurs when these isopycnals intersect the euphotic layers.

Coastal tidal processes

Simpson and Hunter (1974), Pingree and Griffiths (1978), and Pingree et al. (1975) have pioneerd the approach of using satellite

imagery and modelling to the study of tidal frontal phenomena in
the waters around Great Britain. Remote sensing was needed to
obtain information on water mass structure and its pigment distribution and to obtain these parameters in a synoptic fashion over
wide areas. In the final analysis, the concepts derived from either
observation and/or numerical modelling were substantiated or
reinforced by remote sensing capabilities.

In this section, I will describe a similar study which essentially began in 1927 with a series of shipboard observations by H.B.
Bigelow in the area of the Gulf of Maine and Georges Bank. His
conclusions as the result of the observation are confirmed by
satellite imagery taken in 1979. In the beginning, Bigelow measured
the thermal structure of water masses of the Gulf of Maine and
Georges Bank and computed the stability of these water masses to
outline different regimes of vertical mixing. From this analysis,
he concluded that the different regimes of temperature which
outline the areas of vertical mixing were due to the intense tidal
action throughout the area. More recently, Garrett et al. (1978)
subjected this region (Fig. 13) to an analysis using a numerical
model developed by Simpson and Hunter (1974). This model proposes
that the difference between mixed and stratified waters is dictated
by an index or ratio of the potential energy (required to thorou-
ghly mix the water) to the rate of energy that is dissipated by
the flow or tidal current across the bottom. The relevant parameter
of index for separating mixed from stratified waters by tide is
referred to as $\log H/U^3$, where the water, H, is divided by the
tidal velocity frictional component, U^3. Essentially, this numerical model (Fig. 13) confirmed Bigelow's original observations that
tidally mixed areas were centered on Georges Bank and Nantucket
Shoals. It also identified other tidal regions off Nova Scotia and
in the Bay of Fundy. The question now asked is how real the model
is and/or how accurate Bigelow's original observations are -this is
where the satellite images can help us.

Comparison of satellite thermal and colorimetric imagery
(CZCS, Fig. 14) with the numerical model and Bigelow's observations, confirms that much of the mixing is tidally driven. In the
thermal image, the light areas indicate warm water and the dark
areas indicate colder waters. The region of Georges Bank and
Nantucket Shoals clearly shows up as well as the cold tidally
mixed regions off Nova Scotia. Dark filamentous segments appear
to be intrusions of either warm slope water of the Gulf Stream and

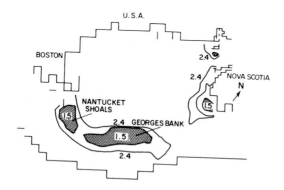

Fig. 13. Numerical model of tidal mixing, log H/U^3 (after Garrett et al., 1978); 1.5 indicates areas totally mixed by tides.

other mixed areas that had not been identified by either observations or modelling.

Satellite thermal imagery compared with the bathymetry of this area gives information with regard to the critical mixed depth for tidal activity. The mixed fronts around Georges Bank appear to center on the 60m isobath; this depth appears to be rather consistent for the entire region.

The significance of tidal mixing on phytoplankton abundance is explained as follows : During the months when the water column is being heated in this region, the greatest buoyancy of surface waters tends to isolate the nutrient rich water from the euphotic zone. Therefore, the restoration of growth by vertical mixing of nutrients into the euphotic zone becomes crucial in regulating the rate of phytoplankton growth. The conceptual model of the density and nutrient distribution across Georges Bank explains why the Bank itself imparts color and temperature signatures on the water (Fig. 15). Nutrient rich water in deeper waters is brought up into the euphotic zone by the tidal action at the frontal edge on either side of the Bank. This water is mixed across the top of the Bank which is in the euphotic zone and promotes luxurious growth on top of the Bank.

In summary, the CZCS colorimetry shows that high concentrations of phytoplankton pigment are located on the Bank and the other frontal regions which outline the areas of vertical mixing, such as Nantucket Shoals. The low phytoplankton pigment concentrations

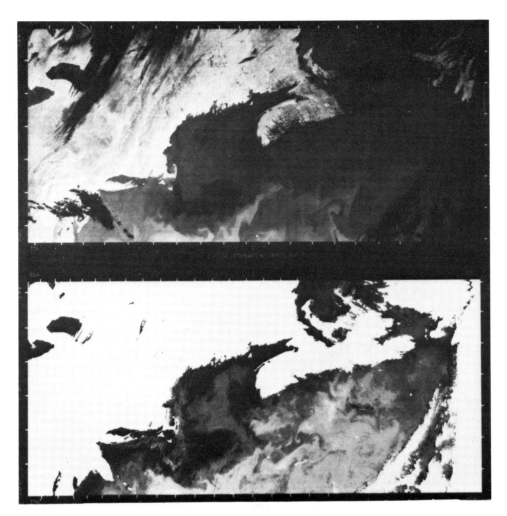

Fig. 14. Orbit 3326 14 June 1979. Top image : sea surface temperature; dark, cold water; light area, warm water. Bottom image :
phytoplankton pigment; dark, high pigment; light, low pigment.
(Yentsch and Garfield, 1981).

occur in the slope waters or in the central region of the Gulf of
Maine where tidal mixing and bottom friction action is not effective.

Passive tracer or growth

There appear to be two obvious hypotheses to attribute to the
patterns of ocean color. 1) Distribution of phytoplankton pigments
are <u>passive tracers</u> to the movements of surface waters and,
2) Distribution of phytoplankton reflects the fluid dynamics of the

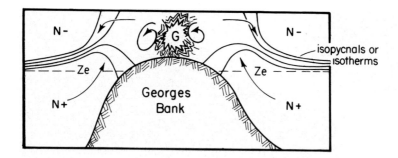

Fig. 15. Conceptual diagram of nutrient enrichment on Georges Bank.

water masses which supplies nutrients for phytoplankton growth - the hypothesis of <u>nutrient enrichment</u> for growth.
The first hypothesis is unattractive because the satellite images that I have observed show a close correlation between the thermal signatures and the colorimetric signatures. If the low nutrient concentrations in the surface waters of the ocean are limiting growth and hence, the abundance of phytoplankton, one would expect that the horizontal diffusion would progressively disperse the phytoplankton. Any correlation between temperature and color would come about almost by accident.

The <u>nutrient enrichment</u> hypothesis argues that it is the vertical flux of nutrients which regulate phytoplankton abundance. It is through this process that one can account for the close correspondence between temperature and water color observed in satellite imagery. This hypothesis also argues that in order to have correspondence between temperature and color, growth must be in excess of that removed by grazing or sinking by the phytoplankton pigment, and is consistent with our concept of how productivity is regulated. In short, regulation of abundance is brought about by periodic injections of nutrients which change the growth rate in the surface waters of the ocean.

The satellite imagery shown in this paper demonstrates that color changes are closely associated with vertical mixing. The CZCS color pigment patterns are not an undecipherable mix, but clearly reflect the role vertical mixing plays in nutrient supply. A paradox arises : the acquisition of buoyancy to water masses is the antithesis to growth. But growth occurs throughout the oceans because certain forces tend to override the buoyant forces. These forces are largely associated with ocean currents and the vertical mixing as a result of bottom friction, and/or the differences between the density of the water masses.

The definition that I have used here of large scale motion
requires better definition. The scale of the motions I am discus-
sing are those that are influenced by earth's rotation - that is
water motions whose Rossby number is characteristically very small -
hence the large scale motions are comparatively slower than the
velocity imposed by the earth's rotation. One imagines that in
water masses where the Rossby number is very large - that is, the
flow is large compared to the earth's rotation - color pigment
relationships could probably be treated in a Lagrangian sense.
Flow of this sort is uncharacteristic of the open ocean.

Summary

At the opening of this symposium, Jacques Nihoul stressed that
remote sensing occupies a companion role with conceptual and
numerical modelling. Both are the principle tools by which oceanographers can study their medium. The numerical models used in this
text and reported elsewhere, document the interrelationship
between modelling and remote sensing. In order for this approach
to be successful the modeller has to acquire a mental picture of
the pattern of events that will occur in the sea and have the
capability of comparing these patterns to a satellite image. As
more satellite imagery becomes available to the biological oceanographer, pattern recognition will become important. This recognition will depend on better measurements of motions in the ocean
interior, as well as an appreciation of the size of these features.

ACKNOWLEDGEMENTS

The author greatly acknowledges the assistance of Pat Boisvert
and Jim Rollins in preparing the manuscript. The work was funded
by the National Aeronautics and Space Administration, the Office
of Naval Research, the National Science Foundation and the State
of Maine.

REFERENCES

Garrett, C.J.R., Keeley, J.R. and Greenberg, D.A., 1978. Tidal
    mixing versus thermal stratification in the Bay of Fundy and
    the Gulf of Maine. Atmosphere-Ocean, 16: 403-423.
Gordon, H.R., Clark, D.K., Brown, J.W., Brown, O.B. and Evans,
    R.H., 1982. Satellite measurements of phytoplankton concen-

tration in the surface waters of a warm core Gulf Stream ring. J. Mar. Res., 40: 491-502.

Nowlin, W.D., Jr. and McLellan, H.J., 1967. A characterization of Gulf of Mexico waters in winter. J. Mar. Res., 25(1): 29-59.

Pingree, R.D. and Griffiths, D.K., 1978. Tidal fronts on the shelf seas around the British Isles. J. Geophys. Res., 83: 4615-4622.

Pingree, R.D., Pugh, P.R., Holligan, P.M. and Forster, G.R., 1975. Summer phytoplankton blooms and red tides along tidal fronts in the approaches to the English Channel. Nature, London, 258: 672-677.

Richardson, P.L., Cheney, R.E. and Worthington, L.V., 1978. A census of Gulf Stream rings, Spring 1975. J. Geophys. Res., 83: 6136-6144.

Rossby, C.G., 1936. Dynamics of steady ocean currents in light of experimental fluid mechanics. Papers in Phys. Oceanogr. and Meteorol., 5(1): 3.

Simpson, J.H. and Hunter, J.R., 1974. Fronts in the Irish Sea. Nature, London, 250: 404-406.

Yentsch, C.S., 1974. The influence of geostrophy on primary production. Tethys, 6(1-2): 111-118.

Yentsch, C.S., 1983. Satellite observation of phytoplankton distribution associated with large scale oceanic circulation. NAFO Sci. Coun. Studies, 4: 53-59.

Yentsch, C.S. and Garfield, N., 1981. Principal areas of vertical mixing in the waters of the Gulf of Maine, with reference to the total productivity of the area. In: J.F.R. Gower (Editor), Oceanography from Space, Plenum Publ. Corp., pp. 303-312.